Geometry

by
Edward Kohn, M.S.

Series Editor
Jerry Bobrow, Ph.D.

CliffS Notes
INCORPORATED
LINCOLN, NEBRASKA   68501

QA
461
.K59
1994

FIRST EDITION

ISBN 0-8220-5328-4

CONTENTS

CONTENTS

Introduction

Geometry was the first system of ideas in which a few simple statements were assumed (**postulates**) and then used to derive more complex ones (**theorems**). A system such as this is referred to as a deductive system. Geometry introduces you to the ideas of deduction and logical consequences, ideas you will continue to use throughout your life.

Points, Lines, and Planes

- **Point.** A **point** is the most fundamental idea in geometry. It is represented by a dot and named by a capital letter. Figure 1 illustrates point C, point M, and point Q.

■ Figure 1 ■

- **Line.** A **line** (*straight line*) consists of an infinite number of points. It continues forever in two opposite directions. It is named by any two points on the line. The symbol ↔ written on top of two letters is used to denote that line. A line may also be named by one small letter (Figure 2).

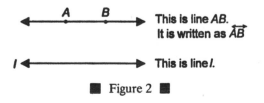

This is line AB.
It is written as $\overleftrightarrow{AB}$

This is line l.

■ Figure 2 ■

- **Collinear points.** Points that lie on the same line are called **collinear points.** If points do *not* lie on the same line, then they are **noncollinear points.** In Figure 3, points M, A, and N are collinear, and points T, I, and c are noncollinear.

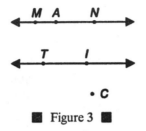

■ Figure 3 ■

- **Plane.** A **plane** is a flat surface which extends indefinitely in all directions. It is usually represented in drawings by a four-sided figure. A single capital letter is used to denote a plane. The word *plane* is written with the letter so as not to be confused with a point (Figure 4).

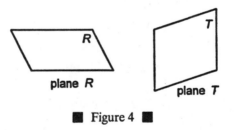

■ Figure 4 ■

- **Postulates and theorems.** As mentioned in the introduction, a **postulate** is a statement assumed true without proof. A **theorem** is a true statement that can be proven. Listed below are the first five postulates and the theorems that can be proven from these postulates.

Postulate 1: A line contains at least two points.

Postulate 2: A plane contains at least three noncollinear points.

Postulate 3: Through any two points, there is exactly one line.

Postulate 4: Through any three noncollinear points, there is exactly one plane.

Postulate 5: If two points lie in a plane, then the line joining them lies in that plane.

Postulate 6: If two planes intersect, then their intersection is a line.

Theorem 1: If two lines intersect, then they intersect in exactly one point.

Theorem 2: If a point lies outside a line, then exactly one plane contains both the line and the point.

Theorem 3: If two lines intersect, then exactly one plane contains both lines.

Example 1: State the postulate or theorem you would use to justify the statement made about each figure.

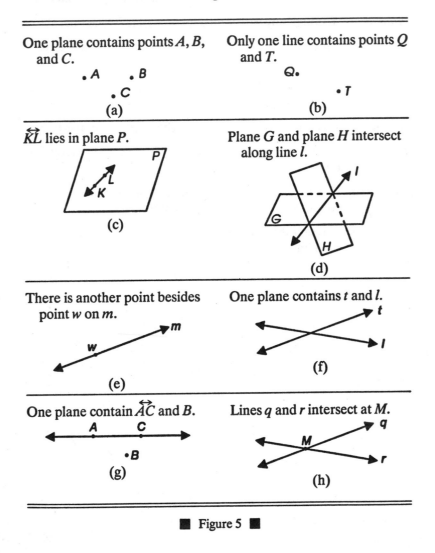

One plane contains points A, B, and C.

. A . B

. C

(a)

Only one line contains points Q and T.

Q.

. T

(b)

$\overleftrightarrow{KL}$ lies in plane P.

(c)

Plane G and plane H intersect along line l.

(d)

There is another point besides point w on m.

(e)

One plane contains t and l.

(f)

One plane contain $\overleftrightarrow{AC}$ and B.

(g)

Lines q and r intersect at M.

(h)

■ Figure 5 ■

(a) Through any three noncollinear points, there is exactly one plane (*Postulate 4*).

(b) Through any two points, there is exactly one line (*Postulate 3*).

(c) If two points lie in a plane, then the line joining them lies in that plane (*Postulate 5*).

(d) If two planes intersect, then their intersection is a line (*Postulate 6*).

(e) A line contains at least two points (*Postulate 1*).

(f) If two lines intersect, then exactly one plane contains both lines (*Theorem 3*).

(g) If a point lies outside a line, then exactly one plane contains both the line and the point (*Theorem 2*).

(h) If two lines intersect, then they intersect in exactly one point (*Theorem 1*).

Segments, Midpoints, and Rays

Line segment. A line segment is a piece of a line. It has two endpoints and is named by its endpoints. Sometimes, the symbol ‾ written on top of two letters is used to denote the segment. This is line segment CD (Figure 6).

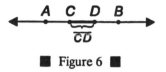

■ Figure 6 ■

It is written $\overline{CD}$ (Technically, $\overline{CD}$ refers to the points C and D and all the points between them, and CD without the ‾ refers to the distance from C to D.). Note that $\overline{CD}$ is a piece of $\overleftrightarrow{AB}$.

Postulate 7: (*Ruler Postulate*) Each point on a line can be paired with exactly one real number called its **coordinate.** The distance between two points is the positive difference of their coordinates (Figure 7).

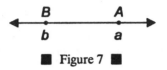

■ Figure 7 ■

If $a > b$, then $AB = a - b$.

Example 2: In Figure 8, find the length of QU.

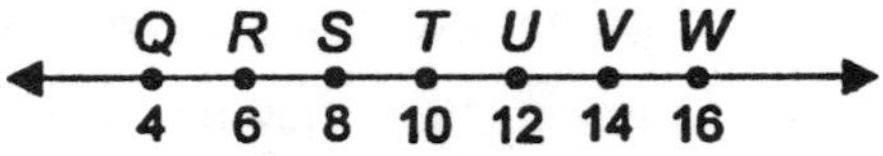

$$
\begin{array}{ccccccc}
Q & R & S & T & U & V & W \\
4 & 6 & 8 & 10 & 12 & 14 & 16
\end{array}
$$

■ Figure 8 ■

$QU = 12 - 4$

$QU = 8$ (The length of $\overline{QU}$ is 8.)

Postulate 8: (*Segment Addition Postulate*) If B lies between A and C on a line, then $AB + BC = AC$ (Figure 9).

A B   C

■ Figure 9 ■

Example 3: In Figure 10, A lies between C and T. Find CT if $CA = 5$ and $AT = 8$.

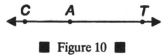

■ Figure 10 ■

Since A lies between C and T, *Postulate 8* tells you

$$CA + AT = CT$$
$$5 + 8 = 13$$
$$CT = 13$$

Midpoint. A **midpoint** of a line segment is the halfway point, or the point equidistant from the endpoints (Figure 11).

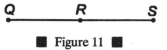

■ Figure 11 ■

R is the midpoint of $\overline{QS}$ because $QR = RS$ or because $QR = \frac{1}{2}QS$ or $RS = \frac{1}{2}QS$.

Example 4: In Figure 12, find the midpoint of $\overline{KR}$.

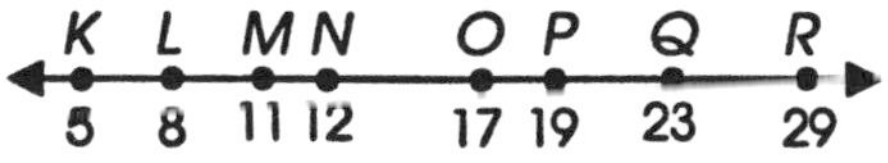

K L M N   O P Q   R
5 8 11 12   17 19 23   29

■ Figure 12 ■

$$KR = 29 - 5 \quad \text{or} \quad KR = 24$$

The midpoint of $\overline{KR}$ would be ½(24) or 12 spaces from either K or R. Since the coordinate of K is 5, and it is smaller than the coordinate of R, which is 29, one way to get the coordinate of the midpoint would be either to add 12 to 5 or to subtract 12 from 29. In either case, you determine that the coordinate of the midpoint is 17. That means that point O is the midpoint of $\overline{KR}$ because $KO = OR$.

Another way to get the coordinate of the midpoint would be to find the average of the endpoint coordinates. To find the average of two numbers, you find their sum and divide by two. (5 +.29) ÷ 2 = 17. The coordinate of the midpoint is 17, so the midpoint is point O.

Theorem 4: A line segment has exactly one midpoint.

Ray. A **ray** is also a piece of a line, except that it has only one endpoint and continues forever in one direction. It could be thought of as a half-line with an endpoint. It is named by the letter of its endpoint and any other point on the ray. The symbol → written on top of the two letters is used to denote that ray. This is ray *AB* (Figure 13).

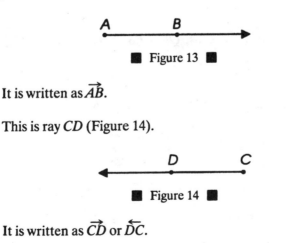

■ Figure 13 ■

It is written as $\overrightarrow{AB}$.

This is ray *CD* (Figure 14).

■ Figure 14 ■

It is written as $\overrightarrow{CD}$ or $\overleftarrow{DC}$.

Note that the nonarrow part of the ray symbol is over the endpoint.

Angles and Angle Pairs

Angle. An **angle** is formed by two rays that have the same endpoint. That point is called the **vertex;** the rays are called the **sides** of the angle. In geometry, an angle is measured in **degrees** from 0° to 180°. The number of degrees indicates the size of the angle. In Figure 15, the angle is formed by rays AB and AC. A is the vertex. $\overrightarrow{AB}$ and $\overrightarrow{AC}$ are the sides of the angle.

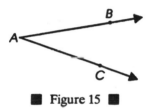

■ Figure 15 ■

The symbol ∠ is used to denote an angle. The symbol $m \angle$ is sometimes used to denote the measure of an angle.

An angle can be named in various ways (Figure 16).

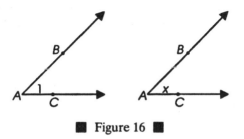

■ Figure 16 ■

1. By the letter of the vertex—therefore, the angle in Figure 16 could be named ∠A.

2. By the number (or small letter) in its interior—therefore, the angle in Figure 16 could be named ∠1 or ∠x.

3. By the letters of three points that form it—therefore, the angle in Figure 16 could be named ∠*BAC* or ∠*CAB*. The center letter is always the letter of the vertex.

Example 5: In Figure 17 (a) use three letters to rename ∠3; (b) use one number to rename ∠*KMJ*.

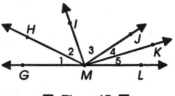

■ Figure 17 ■

(a) ∠3 is the same as ∠*IMJ* or ∠*JMI*;

(b) ∠*KMJ* is the same as ∠4.

Postulate 9: (*Protractor Postulate*) Suppose *O* is a point of $\overleftrightarrow{XY}$. Consider all rays with endpoint *O* which lie on one side of $\overleftrightarrow{XY}$. Each ray can be paired with exactly one real number between 0° and 180°, as shown in Figure 18. The positive difference between two numbers representing two different rays is the measure of the angle.

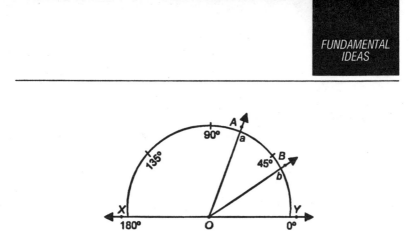

■ Figure 18 ■

If $a > b$, then $m \angle AOB = a - b$.

Example 6: Use Figure 19 to find the following: (a) $m \angle SON$, (b) $m \angle ROT$, (c) $m \angle MOE$.

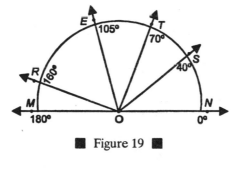

■ Figure 19 ■

(a) $m \angle SON = 40° - 0°$ (b) $m \angle ROT = 160° - 70°$

  $m \angle SON = 40°$ $m \angle ROT = 90°$

(c) $m \angle MOE = 180° - 105°$

  $m \angle MOE = 75°$

Postulate 10: (*Angle Addition Postulate*) If $\overrightarrow{OB}$ lies between $\overrightarrow{OA}$ and $\overrightarrow{OC}$, then $m \angle AOB + m \angle BOC = m \angle AOC$ (Figure 20).

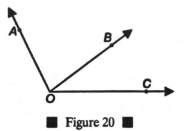

■ Figure 20 ■

Example 7: In Figure 21, if $m \angle 1 = 32°$ and $m \angle 2 = 45°$, find $m \angle NEC$.

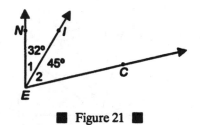

■ Figure 21 ■

Since $\overrightarrow{EI}$ is between $\overrightarrow{EN}$ and $\overrightarrow{EC}$, by *Postulate 10,*

$$m \angle 1 + m \angle 2 = m \angle NEC$$
$$32° + 45° = m \angle NEC$$
$$m \angle NEC = 77°$$

Angle bisector. An **angle bisector** is a ray that divides an angle into two equal angles. In Figure 22, $\overrightarrow{OY}$ is a bisector of $\angle XOZ$ because $m \angle XOY = m \angle YOZ$.

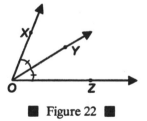

■ Figure 22 ■

Theorem 5: An angle has exactly one bisector.

Certain angles are given special names based on their measures.

Right angle. A **right angle** has a measure of 90°. The symbol ∟ in the interior of an angle designates the fact that a right angle is formed. In Figure 23, ∠*ABC* is a right angle.

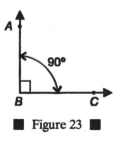

■ Figure 23 ■

$$m \angle ABC = 90°$$

Theorem 6: All right angles are equal.

Acute angle. An **acute angle** is any angle whose measure is less than 90°. In Figure 24, ∠*b* is acute.

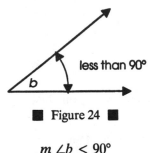

■ Figure 24 ■

$$m \angle b < 90°$$

Obtuse angle. An **obtuse angle** is an angle whose measure is more than 90° but less than 180°. In Figure 25, ∠4 is obtuse.

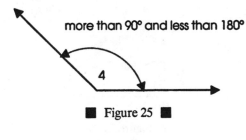

■ Figure 25 ■

$$m \angle 4 > 90° \text{ and } m \angle 4 < 180°$$
$$\text{or}$$
$$90° < m \angle 4 < 180°$$

Straight angle. Some geometry texts refer to an angle with a measure of 180° as a **straight angle**. In Figure 26, ∠BAC is a straight angle.

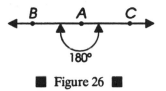

■ Figure 26 ■

$$m \angle BAC = 180°$$

Example 8: Use Figure 27 to identify each named angle as either acute, right, obtuse, or straight: (a) ∠BFD, (b) ∠AFE, (c) ∠BFC, (d) ∠DFA.

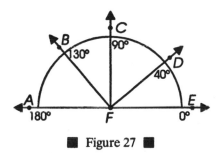

■ Figure 27 ■

(a) $m \angle BFD = 90°$ (130° − 40° = 90°), so ∠BFD is a right angle.

(b) $m \angle AFE = 180°$, so ∠AFE is a straight angle.

(c) $m \angle BFC = 40°$ (130° − 90° = 40°), so ∠BFC is an acute angle.

(d) $m \angle DFA = 140°$ (180° − 40° = 140°), so ∠DFA is an obtuse angle.

Certain angle pairs are given special names based on their relative position to one another or based on the sum of their respective measures.

Adjacent angles. **Adjacent angles** are any two angles that share a common side separating the two angles and that share a common vertex. In Figure 28, ∠1 and ∠2 are adjacent angles.

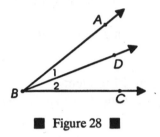

■ Figure 28 ■

Vertical angles. **Vertical angles** are formed when two lines intersect and form four angles. Any two of these angles which are *not* adjacent angles are called vertical angles. In Figure 29, line *l* and line *m* intersect at point *Q* forming ∠1, ∠2, ∠3, and ∠4.

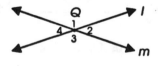

■ Figure 29 ■

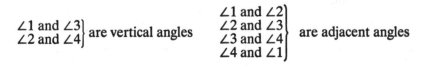

Theorem 7: Vertical angles are equal in measure.

Complementary angles. Complementary angles are any two angles whose sum is 90°. In Figure 30, since $\angle ABC$ is a right angle, $m \angle 1 + m \angle 2 = 90°$, so $\angle 1$ and $\angle 2$ are complementary.

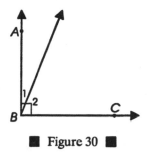

■ Figure 30 ■

Complementary angles do not need to be adjacent. In Figure 31, since $m \angle 3 + m \angle 4 = 90°$, $\angle 3$ and $\angle 4$ are complementary.

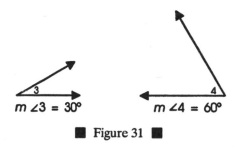

$m \angle 3 = 30°$ $m \angle 4 = 60°$

■ Figure 31 ■

Example 9: If $\angle 5$ and $\angle 6$ are complementary, and $m \angle 5 = 15°$, find $m \angle 6$.

Since $\angle 5$ and $\angle 6$ are complementary,

$$m \angle 5 + m \angle 6 = 90°$$
$$15 + m \angle 6 = 90°$$
$$m \angle 6 = 90° - 15°$$
$$m \angle 6 = 75°$$

Theorem 8: If two angles are complementary to the same angle, or to equal angles, then they are equal to each other.

Refer to Figures 32 and 33. In Figure 32, $\angle A$ and $\angle B$ are complementary. Also, $\angle C$ and $\angle B$ are complementary. *Theorem 8* tells you that $m \angle A = m \angle C$. In Figure 33, $\angle A$ and $\angle B$ are complementary. Also, $\angle C$ and $\angle D$ are complementary, and $m \angle B = m \angle D$. *Theorem 8* now tells you that $m \angle A = m \angle C$.

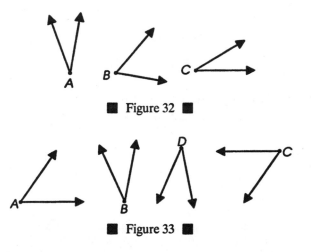

■ Figure 32 ■

■ Figure 33 ■

Supplementary angles. **Supplementary angles** are two angles whose sum is 180°. In Figure 34, $\angle ABC$ is a straight angle. Therefore, $m \angle 6 + m \angle 7 = 180°$, so $\angle 6$ and $\angle 7$ are supplementary.

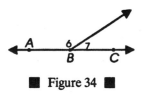

■ Figure 34 ■

Theorem 9: If two adjacent angles lie on a line, then they are supplementary angles.

Supplementary angles do not need to be adjacent (Figure 35).

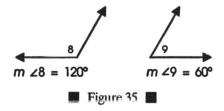

$m \angle 8 = 120°$ $m \angle 9 = 60°$

■ Figure 35 ■

Since $m \angle 8 + m \angle 9 = 180°$, $\angle 8$ and $\angle 9$ are supplementary.

Theorem 10: If two angles are supplementary to the same angle, or to equal angles, then they are equal to each other.

Intersecting, Perpendicular, and Parallel Lines

Intersecting lines. Two or more lines that meet at a point are called **intersecting lines.** That point would be on each of these lines. In Figure 36, lines *l* and *m* intersect at *Q*.

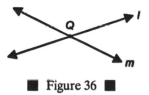

■ Figure 36 ■

Perpendicular lines. Two lines that inersect and form right angles are called **perpendicular lines.** The symbol ⊥ is used to denote perpendicular lines. In Figure 37, line l ⊥ line m.

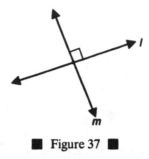

■ Figure 37 ■

Parallel lines. Two lines, both in the same plane, that never intersect are called **parallel lines.** Parallel lines remain the same distance apart at all times. The symbol ∥ is used to denote parallel lines. In Figure 38, $l \parallel m$.

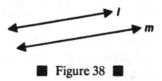

■ Figure 38 ■

Parallel and Perpendicular Planes

Parallel planes. **Parallel planes** are two planes that do not intersect. In Figure 39, plane $P \parallel$ plane Q.

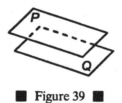

■ Figure 39 ■

Theorem 11: If two planes are parallel to a third plane, then the two planes are parallel to each other (Figure 40).

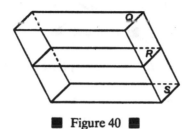

■ Figure 40 ■

Plane Q ∥ plane R, plane S ∥ plane R;

thus, plane Q ∥ plane S.

Perpendicular planes. Perpendicular planes are two planes that intersect so that one of them contains a line perpendicular to the other. In Figure 41, $\overleftrightarrow{AB}$ lies in plane S, and $\overleftrightarrow{AB}$ ⊥ plane T, so plane S ⊥ plane T.

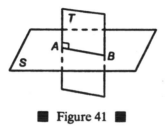

■ Figure 41 ■

Theorem 12: **If two planes are penpendicular to the same plane,
then the two planes either intersect or are parallel. In
Figure 42, plane $B \perp$ plane A, plane $C \perp$ plane A, and
plane B and plane C intersect along line l.**

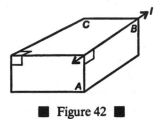

■ Figure 42 ■

In Figure 43, plane $B \perp$ plane A, plane $C \perp$ plane A, and plane
$B \parallel$ plane C.

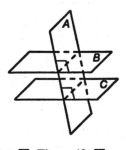

■ Figure 43 ■

Angle Pairs Created with a Transversal

A **transversal** is any line that intersects two or more lines in the same plane but at different points. In Figure 44, line *t* is a transversal.

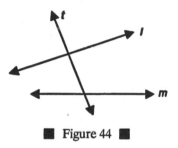

■ Figure 44 ■

A transversal that intersects two lines forms eight angles; certain pairs of these angles are given special names.

- **Corresponding angles** are the angles that appear to be in the same relative position in each group of four angles. In Figure 45, ∠1 and ∠5 are corresponding angles. Other pairs of corresponding angles in Figure 45 are: ∠4 and ∠8, ∠2 and ∠6, and ∠3 and ∠7.

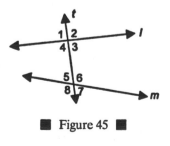

■ Figure 45 ■

- **Alternate interior angles** are angles within the lines being intersected, on opposite sides of the transversal, and are not adjacent. In Figure 45, ∠4 and ∠6 are alternate interior angles. Also, ∠3 and ∠5 are alternate interior angles.

- **Alternate exterior angles** are angles outside the lines being intersected, on opposite sides of the transversal, and are not adjacent. In Figure 45, ∠1 and ∠7 are alternate exterior angles. Also, ∠2 and ∠8 are alternate exterior angles.

- **Consecutive interior angles** (same-side interior angles) are interior angles on the same side of the transversal. In Figure 45, ∠4 and ∠5 are consecutive interior angles. Also, ∠3 and ∠6 are consecutive interior angles.

- **Consecutive exterior angles** (same-side exterior angles) are exterior angles on the same side of the transversal. In Figure 45, ∠1 and ∠8 are consecutive exterior angles. Also, ∠2 and ∠7 are consecutive exterior angles.

Parallel Postulate

Postulate 11: If two parallel lines are cut by a transversal, then the corresponding angles are equal (Figure 46).

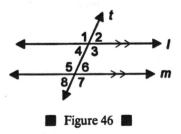

■ Figure 46 ■

This postulate says that if $l \parallel m$, then
$$m \angle 1 = m \angle 5$$
$$m \angle 2 = m \angle 6$$
$$m \angle 3 = m \angle 7$$
$$m \angle 4 = m \angle 8$$

Consequences of the Parallel Postulate

Postulate 11 can be used to derive additional theorems regarding parallel lines cut by a transversal. In Figure 46,

since $\left. \begin{array}{l} m \angle 1 + m \angle 2 = 180° \\ m \angle 5 + m \angle 6 = 180° \end{array} \right\}$ because adjacent angles on a line are supplementary

and $\left. \begin{array}{l} m \angle 1 = m \angle 3 \\ m \angle 2 = m \angle 4 \\ m \angle 5 = m \angle 7 \\ m \angle 6 = m \angle 8 \end{array} \right\}$ because vertical angles are equal

all of the following theorems can be proven as a consequence of *Postulate 11*.

Theorem 13: If two parallel lines are cut by a transversal, then alternate interior angles are equal.

Theorem 14: If two parallel lines are cut by a transversal, then alternate exterior angles are equal.

Theorem 15: If two parallel lines are cut by a transversal, then consecutive interior angles are supplementary.

Theorem 16: If two parallel lines are cut by a transversal, then consecutive exterior angles are supplementary.

The above postulate and theorems can be condensed to the following theorems:

Theorem 17: If two parallel lines are cut by a transversal, then every pair of angles formed are either equal or supplementary.

Theorem 18: If a transversal is perpendicular to one of two parallel lines, then it is also perpendicular to the other line.

Based on *Postulate 11* and the theorems that follow it, any of the following conditions would be true if $l \parallel m$ (Figure 47).

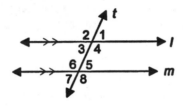

■ Figure 47 ■

$m \angle 1 = m \angle 5$
$m \angle 4 = m \angle 8$
$m \angle 2 = m \angle 6$ based on *Postulate 11*
$m \angle 3 = m \angle 7$

$m \angle 3 = m \angle 5$
$m \angle 4 = m \angle 6$ based on *Theorem 13*

$m \angle 1 = m \angle 7$
$m \angle 2 = m \angle 8$ based on *Theorem 14*

$\angle 3$ and $\angle 6$ are supplementary
$\angle 4$ and $\angle 5$ are supplementary based on *Theorem 15*

∠1 and ∠8 are supplementary⎤
∠2 and ∠7 are supplementary⎦ based on *Theorem 16*

If $t \perp l$, then $t \perp m$.⎦ based on *Theorem 18*

Testing for Parallel Lines

Many times in geometry you will be asked to prove that a pair of lines are parallel. In order to do this, you must look for the converse (reverse) of what was true if the lines had already been parallel.

Postulate 12: If two lines and a transversal form equal corresponding angles, then the lines are parallel.

In Figure 48, if $m \angle 1 = m \angle 2$, then $l \parallel m$. (Any pair of equal corresponding angles would make $l \parallel m$.)

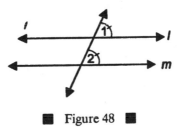

■ Figure 48 ■

This postulate allows you to prove that all the converses (reverses) of the previous theorems are also true.

Theorem 19: If two lines and a transversal form equal alternate interior angles, then the lines are parallel.

Theorem 20: If two lines and a transversal form equal alternate exterior angles, then the lines are parallel.

Theorem 21: If two lines and a transversal form consecutive interior angles that are supplementary, then the lines are parallel.

Theorem 22: If two lines and a transversal form consecutive exterior angles that are supplementary, then the lines are parallel.

Theorem 23: In a plane, if two lines are parallel to a third line, then the two lines are parallel.

Theorem 24: In a plane, if two lines are perpendicular to the same line, then the two lines are parallel.

Based on *Postulate 12* and the theorems that follow it, any of the following conditions would allow you to prove that $a \parallel b$ (Figure 49).

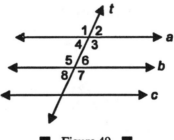

■ Figure 49 ■

$$\left. \begin{array}{l} m \angle 1 = m \angle 5 \\ m \angle 2 = m \angle 6 \\ m \angle 3 = m \angle 7 \\ m \angle 4 = m \angle 8 \end{array} \right\} \text{ use } Postulate\ 12$$

$$\left. \begin{array}{l} m \angle 4 = m \angle 6 \\ m \angle 3 = m \angle 5 \end{array} \right\} \text{ use } Theorem\ 19$$

$$\left. \begin{array}{l} m \angle 1 = m \angle 7 \\ m \angle 2 = m \angle 8 \end{array} \right\} \text{ use } Theorem\ 20$$

∠4 and ∠5 are supplementary⎱
∠3 and ∠6 are supplementary⎰ use *Theorem 21*

∠1 and ∠8 are supplementary⎱
∠2 and ∠7 are supplementary⎰ use *Theorem 22*

$a \parallel c$ and $b \parallel c$} use *Theorem 23*

$a \perp t$ and $b \perp t$} use *Theorem 24*

Example 1: Using Figure 50, identify the given angle pairs as alternate interior, alternate exterior, consecutive interior, consecutive exterior, corresponding, or none of these: ∠1 and ∠7, ∠2 and ∠8, ∠3 and ∠4, ∠4 and ∠8, ∠3 and ∠2, ∠5 and ∠7.

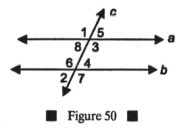

■ Figure 50 ■

∠1 and ∠7 are alternate exterior angles.

∠2 and ∠8 are corresponding angles.

∠3 and ∠4 are consecutive interior angles.

∠4 and ∠8 are alternate interior angles.

∠3 and ∠2 are none of these.

∠5 and ∠7 are consecutive exterior angles.

Example 2: For each of the following figures, determine which postulate or theorem you would use to prove $l \parallel m$.

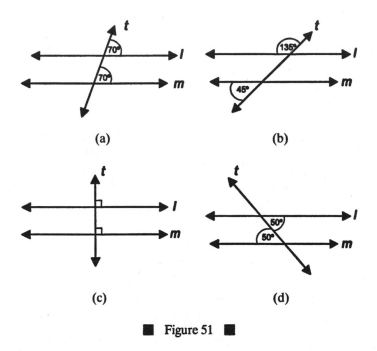

(a)   (b)

(c)   (d)

■ Figure 51 ■

Figure 51(a)—If two lines and a transversal form equal corresponding angles, then the lines are parallel (*Postulate 12*).

Figure 51(b)—If two lines and a transversal form consecutive exterior angles that are supplementary, then the lines are parallel (*Theorem 22*).

Figure 51(c)—In a plane, if two lines are perpendicular to the same line, the the two lines are parallel (*Theorem 24*).

Figure 51(d)—If two lines and a transversal form equal alternate interior angles, then the lines are parallel (*Theorem 19*).

Example 3: In Figure 52, $a \parallel b$ and $m \angle 1 = 117°$. Find the measure of each of the numbered angles.

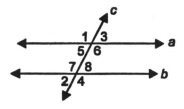

■ Figure 52 ■

$m \angle 2 = 63°$ $m \angle 6 = 117°$
$m \angle 3 = 63°$ $m \angle 7 = 117°$
$m \angle 4 = 117°$ $m \angle 8 = 63°$
$m \angle 5 = 63°$

A **triangle** is a three-sided figure. It has three angles in its interior. The symbol for triangle is △. A triangle is named by the three letters in its vertices. This is △ *ABC* (Figure 53).

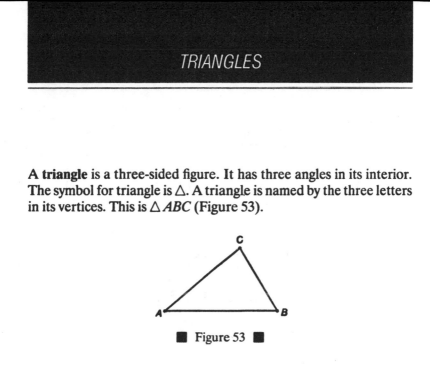

■ Figure 53 ■

Angle Sum of a Triangle

With the use of the *Parallel Postulate,* the following theorem can be proven.

Theorem 25: The sum of the interior angles of any triangle is 180°.

In Figure 53, $m \angle A + m \angle B + m \angle C = 180°$.

Example 1: If $m \angle A = 40°$ and $m \angle B = 60°$, find $m \angle C$.

Since $\quad m \angle A + m \angle B + m \angle C = 180°$

then $\qquad\qquad\qquad m \angle C = 180° - (m \angle A + m \angle B)$

$\qquad\qquad\qquad\qquad m \angle C = 180° - (40° + 60°)$

$\qquad\qquad\qquad\qquad m \angle C = 80°$

Exterior Angle of a Triangle

An **exterior angle of a triangle** is formed when one side of a triangle is extended. The angle outside the triangle, but adjacent to an interior angle, is an exterior angle of the triangle (Figure 54).

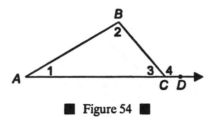

■ Figure 54 ■

In Figure 54, $\angle BCD$ is an exterior angle of $\triangle ABC$.

Since $m \angle 1 + m \angle 2 + m \angle 3 = 180°$ and $m \angle 3 + m \angle 4 = 180°$, you can prove that $m \angle 4 = m \angle 1 + m \angle 2$. This is stated as a theorem.

Theorem 26: An exterior angle of a triangle is equal to the sum of the two **remote** (nonadjacent) interior angles.

Example 2: In figure 54, if $m \angle 1 = 30°$ and $m \angle 2 = 100°$, find $m \angle 4$.

Since $\angle 4$ is an exterior angle of the triangle,

$$m \angle 4 = m \angle 1 + m \angle 2$$
$$m \angle 4 = 30° + 100°$$
$$m \angle 4 = 130°$$

Classifying Triangles

Triangles can be classified either according to their sides or according to their angles.

Types of triangles by sides

- **Equilateral triangle. An equilateral triangle** is a triangle with all three sides equal in measure. In Figure 55, the slash marks indicate equal measure.

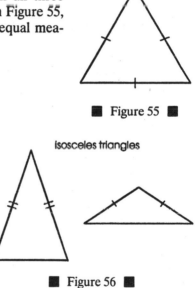

equilateral triangle

■ Figure 55 ■

- **Isosceles triangle. An isosceles triangle** is a triangle in which at least two sides have equal measure (Figure 56).

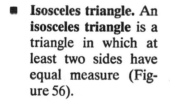

isosceles triangles

■ Figure 56 ■

- **Scalene triangle.** A scalene triangle is a triangle with all three sides of different measures (Figure 57).

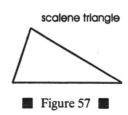

scalene triangle

■ Figure 57 ■

Types of triangles by angles

- **Right triangle.** A **right triangle** is a triangle that has a right angle in its interior (Figure 58).

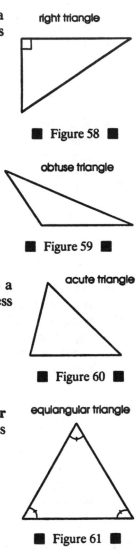

right triangle

■ Figure 58 ■

- **Obtuse triangle.** An **obtuse triangle** is a triangle having an obtuse angle (greater than 90° but less than 180°) in its interior. Figure 59 is an obtuse triangle.

obtuse triangle

■ Figure 59 ■

- **Acute triangle.** An **acute triangle** is a triangle having all acute angles (less than 90°) in its interior (Figure 60).

acute triangle

■ Figure 60 ■

- **Equiangular triangle.** An **equiangular triangle** is a triangle having all angles of equal measure (Figure 61).

equiangular triangle

■ Figure 61 ■

Since the sum of all the angles of a triangle is 180°, the following theorem is easily shown.

Theorem 27: Each angle of an equiangular triangle has a measure of 60°.

Special names for sides and angles

- **Legs, base, vertex angle, and base angles.** In an isosceles triangle, the two equal sides are called **legs,** and the third side is called the **base.** The angle formed by the two equal sides is called the **vertex angle.** The other two angles are called **base angles** (Figure 62).

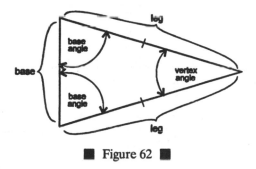

■ Figure 62 ■

- **Hypotenuse and legs.** In a right triangle, the side opposite the right angle is called the **hypotenuse,** and the other two sides are called **legs** (Figure 63).

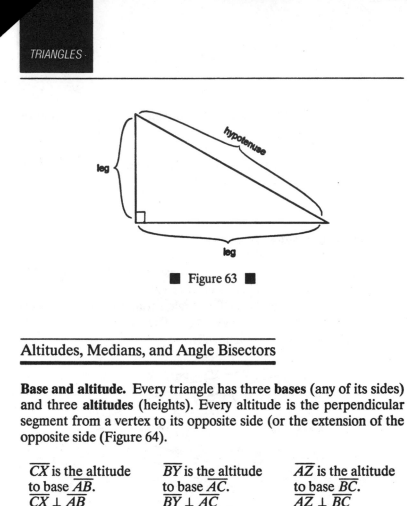

■ Figure 63 ■

Altitudes, Medians, and Angle Bisectors

Base and altitude. Every triangle has three **bases** (any of its sides) and three **altitudes** (heights). Every altitude is the perpendicular segment from a vertex to its opposite side (or the extension of the opposite side (Figure 64).

$\overline{CX}$ is the altitude to base $\overline{AB}$.
$\overline{CX} \perp \overline{AB}$

$\overline{BY}$ is the altitude to base $\overline{AC}$.
$\overline{BY} \perp \overline{AC}$

$\overline{AZ}$ is the altitude to base $\overline{BC}$.
$\overline{AZ} \perp \overline{BC}$

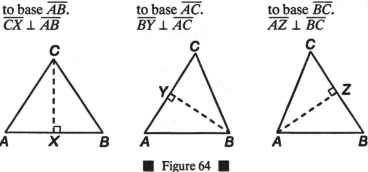

■ Figure 64 ■

Altitudes can sometimes coincide with a side of the triangle and sometimes will meet outside the triangle. In Figure 65, $\overline{AC}$ is an altitude to base $\overline{BC}$ and $\overline{BC}$ is an altitude to base $\overline{AC}$.

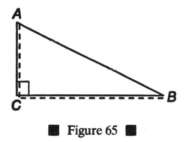

■ Figure 65 ■

In Figure 66, $\overline{AM}$ is the altitude to base $\overline{BC}$.

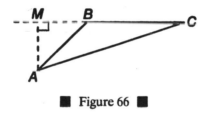

■ Figure 66 ■

It is an interesting fact that in any triangle, the three lines containing the altitudes meet in one point (Figure 67).

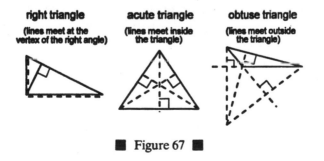

■ Figure 67 ■

Median. The **median** in a triangle is the line segment drawn from a vertex to the midpoint of its opposite side. Every triangle has three medians. In Figure 68, E is the midpoint of $\overline{BC}$. Therefore, $BE = EC$. $\overline{AE}$ is a median of $\triangle ABC$.

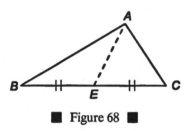

■ Figure 68 ■

In every triangle, the three medians meet in one point inside the triangle (Figure 69).

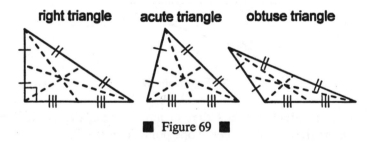

right triangle   acute triangle   obtuse triangle

■ Figure 69 ■

Angle bisector. An **angle bisector** in a triangle is a segment drawn from a vertex and which bisects (cuts in half) that vertex angle. Every triangle has three angle bisectors. In Figure 70, $\overline{BX}$ is an angle bisector in $\triangle ABC$.

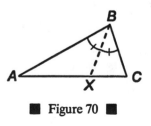

■ Figure 70 ■

In every triangle, the three angle bisectors meet in one point inside the triangle (Figure 71).

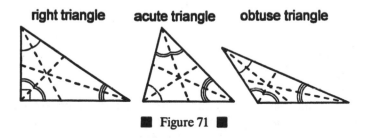

right triangle   **acute triangle**   **obtuse triangle**

■ Figure 71 ■

In general, altitudes, medians, and angle bisectors are different segments. In certain triangles, though, they can be the same segments. In Figure 72, the altitude drawn from the vertex angle of an isosceles triangle can be proven to also be a median, as well as an angle bisector.

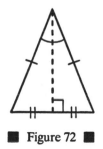

■ Figure 72 ■

Example 3: In Figure 73, based on the markings, name an altitude of $\triangle QRS$; name a median of $\triangle QRS$; and name an angle bisector of $\triangle QRS$.

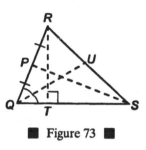

■ Figure 73 ■

$\overline{RT}$ is an altitude to base $\overline{QS}$ because $\overline{RT} \perp \overline{QS}$.

$\overline{SP}$ is a median to base $\overline{QR}$ because P is the midpoint of $\overline{QR}$.

$\overline{QU}$ is an angle bisector of $\triangle QRS$ because it bisects $\angle RQS$.

Congruent Triangles

Triangles that have exactly the same size and shape are called **congruent triangles.** The symbol for congruent is ≅. Two triangles are congruent when the three sides and the three angles of one triangle have the same measurements as three sides and three angles of another triangle. The triangles in Figure 74 are **congruent** triangles.

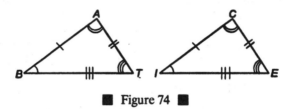

■ Figure 74 ■

Corresponding parts. The parts of the two triangles that have the same measurements are referred to as **corresponding parts.** This means that *Corresponding Parts of Congruent Triangles are Equal (CPCTE).* Congruent triangles are named by listing their vertices in corresponding orders. In Figure 74, $\triangle BAT \cong \triangle ICE$.

Example 4: If $\triangle PQR \cong \triangle STU$, which parts must have equal measurements?

$$m \angle P = m \angle S \qquad PQ = ST$$
$$m \angle Q = m \angle T \qquad QR = TU$$
$$m \angle R = m \angle U \qquad PR = SU$$

because corresponding parts of congruent triangles are equal.

Tests for congruence. To show that two triangles are congruent, it is not necessary to show that all six pairs of corresponding parts are equal. The following postulates and theorems are the most common methods for proving triangles congruent.

Postulate 13: (*SSS Postulate*) If each side of one triangle is congruent to the corresponding side of another triangle, then the triangles are congruent.

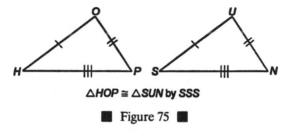

△*HOP* ≅ △*SUN* by *SSS*

■ Figure 75 ■

Postulate 14: (*SAS Postulate*) If two sides and the angle between them in one triangle are congruent to the corresponding parts in another triangle, then the triangles are congruent.

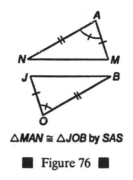

△*MAN* ≅ △*JOB* by *SAS*

■ Figure 76 ■

Postulate 15: (*ASA Postulate*) If two angles and the side between them in one triangle are congruent to the corresponding parts in another triangle, then the triangles are congruent.

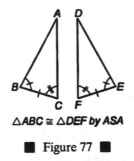

△*ABC* ≅ △*DEF by ASA*

■ Figure 77 ■

Theorem 28: (*AAS Theorem*) If two angles and a side not between them in one triangle are congruent to the corresponding parts in another triangle, then the triangles are congruent.

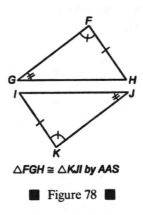

△*FGH* ≅ △*KJI by AAS*

■ Figure 78 ■

Postulate 16: (*HL Postulate*) If the hypotenuse and leg of one right triangle are congruent to the corresponding parts of another right triangle, then the triangles are congruent.

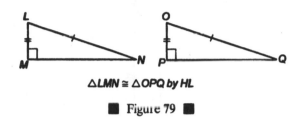

△LMN ≅ △OPQ by HL

■ Figure 79 ■

Theorem 29: (*HA Theorem*) If the hypotenuse and an acute angle of one right triangle are congruent to the corresponding parts of another right triangle, then the triangles are congruent.

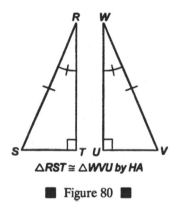

△RST ≅ △WVU by HA

■ Figure 80 ■

Theorem 30: (*LL Theorem*) If the legs of one right triangle are congruent to the corresponding parts of another right triangle, then the triangles are congruent.

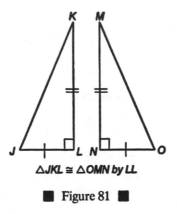

△*JKL* ≅ △*OMN by LL*

■ Figure 81 ■

Theorem 31: (*LA Theorem*) If one leg and an acute angle of one right triangle are congruent to the corresponding parts of another right triangle, then the triangles are congruent.

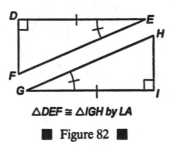

△*DEF* ≅ △*IGH by LA*

■ Figure 82 ■

Example 5: Based on the markings in Figure 83, complete the congruence statement $\triangle ABC \cong \triangle\_$.

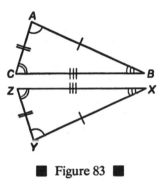

■ Figure 83 ■

$\triangle YXZ$, since A corresponds to Y, B corresponds to X, and C corresponds to Z.

Example 6: By what method would each of the triangles in Figures 84(a) through 84(i) be proven congruent?

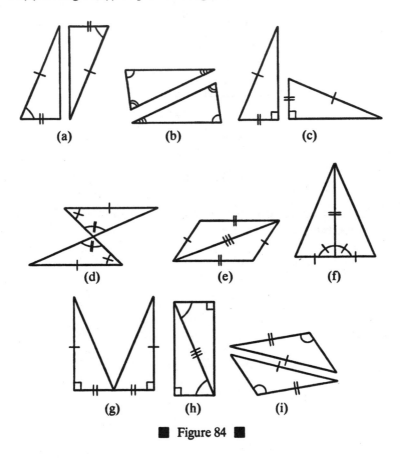

■ Figure 84 ■

(a) *SAS.*

(b) None. There is no *AAA* method.

(c) *HL.*

(d) *AAS.*

(e) *SSS*. The third pair of equal sides is the side that is shared by each triangle.

(f) *SAS*.

(g) *LL* or *SAS*.

(h) *HA* or *AAS*.

(i) None. There is no *SSA* method.

Example 7: Name the additional equal corresponding part(s) needed to prove the triangles in Figures 85(a) through 85(f) congruent by the indicated postulate or theorem.

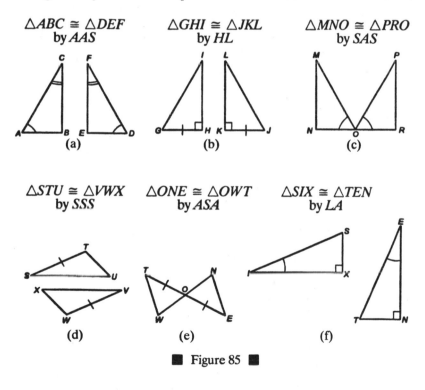

■ Figure 85 ■

(a) $BC = EF$ or $AB = DE$, <u>but not</u> $AC = DF$, since these sides lie between the equal angles.

(b) $GI = JL$.

(c) $MO = PO$ <u>and</u> $NO = RO$.

(d) $TU = WX$ <u>and</u> $SU = VX$.

(e) $m \angle T = m \angle E$ <u>and</u> $m \angle TOW = m \angle EON$.

(f) $IX = EN$ or $SX = TN$, <u>but not</u> $IS = ET$, since they are hypotenuses.

More on Isosceles Triangles

Consider isosceles triangle ABC (Figure 86).

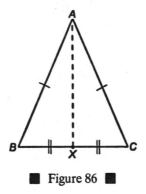

■ Figure 86 ■

With a median drawn from the vertex to the base, $\overline{BC}$, it can be proven that $\triangle BAX \cong \triangle CAX$, which leads to several important theorems.

Theorem 32: If two sides of a triangle are equal, then the angles opposite those sides are also equal.

Theorem 33: If a triangle is equilateral, then it is also equiangular.

Theorem 34: If two angles of a triangle are equal, then the sides opposite these angles are also equal.

Theorem 35: If a triangle is equiangular, then it is also equilateral.

Example 8: Figure 87 has $\triangle QRS$ with $QR = QS$. If $m \angle Q = 50°$, find $m \angle R$ and $m \angle S$.

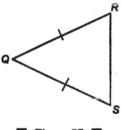

■ Figure 87 ■

Since $m \angle Q + m \angle R + m \angle S = 180°$, and since $QR = QS$ implies that $m \angle R = m \angle S$,

$$m \angle Q + m \angle R + m \angle R = 180°$$

$$50° + 2m \angle R = 180°$$

$$2m \angle R = 130°$$

$$m \angle R = 65° \quad \text{and} \quad m \angle S = 65°$$

Example 9: Figure 88 has $\triangle ABC$ with $m \angle A = m \angle B = m \angle C$, and $AB = 6$. Find BC and AC.

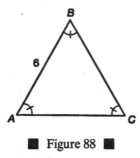

■ Figure 88 ■

Since the triangle is equiangular, it is also equilateral. Therefore, $BC = AC = 6$.

Triangle Inequalities Regarding Sides and Angles

You have just seen that if a triangle has equal sides, the angles opposite these sides are equal, and if a triangle has equal angles, the sides opposite these angles are equal. There are two important theorems involving unequal sides and unequal angles in triangles.

Theorem 36: If two sides of a triangle are unequal, then the measures of the angles opposite these sides are unequal, and the greater angle is opposite the greater side.

Theorem 37: If two angles of a triangle are unequal, then the measures of the sides opposite these angles are also unequal, and the longer side is opposite the greater angle.

Example 10: Figure 89 shows a triangle with angles of different measures. List the sides of this triangle in order from least to greatest.

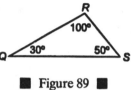

■ Figure 89 ■

Since 30° < 50° < 100°, then $RS < QR < QS$.

Example 11: Figure 90 shows a triangle with sides of different measures. List the angles of this triangle in order from least to greatest.

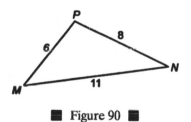

■ Figure 90 ■

Since 6 < 8 < 11, then $m \angle N < m \angle M < m \angle P$.

Example 12: Figure 91 shows right $\triangle ABC$. Which side must be the longest?

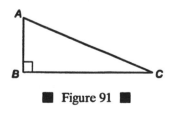

■ Figure 91 ■

Since $\angle B$ is the angle of greatest measure in the triangle, its opposite side is the longest. Therefore, the hypotenuse, $\overline{AC}$, is the longest side in a right triangle.

The Triangle Inequality Theorem

In $\triangle TAB$, Figure 92, if T, A, and B represent three points on a map and you wish to go from T to B, it would be obvious that going from T to A to B would be longer than going directly from T to B. This idea is expressed in the following theorem.

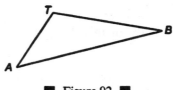

■ Figure 92 ■

Theorem 38: (*Triangle Inequality Theorem*) The sum of the lengths of any two sides of a triangle is greater than the length of the third side.

Example 13: In Figure 93, the measures of two sides of a triangle are 7 and 12. Find the range of possibilities for the third side.

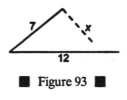

■ Figure 93 ■

Using the *Triangle Inequality Theorem,* you can write the following:

$$7 + x > 12 \quad \text{so} \quad x > 5$$
$$7 + 12 > x \quad \text{so} \quad 19 > x \,(\text{or}\, x < 19)$$

Therefore, the third side must be more than 5 and less than 19.

Closed shapes or figures in a plane with three or more sides are called **polygons.** The term *polygon* is derived from a Greek word meaning "many-angled."

Classifying Polygons

- **Convex and not convex.** Polygons first fit into two general categories—**convex** and **not convex** (sometimes called **concave**). Figure 94 shows some convex polygons, some not-convex polygons, and some figures that are not classified as polygons.

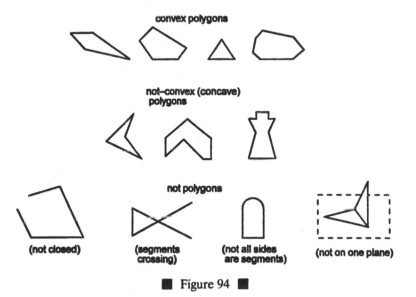

■ Figure 94 ■

- **Vertices.** The endpoints of the sides of polygons are called **vertices.** When naming a polygon, its vertices are named in consecutive order either clockwise or counterclockwise.

- **Consecutive sides. Consecutive sides** refer to two sides that have an endpoint in common. The four-sided polygon in Figure 95 could have been named *ABCD, BCDA,* or *ADCB.* It does not matter with which letter you begin as long as the vertices are named consecutively. Sides $\overline{AB}$ and $\overline{BC}$ are examples of consecutive sides.

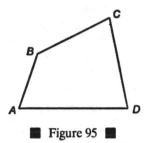

■ Figure 95 ■

- **Diagonal.** A **diagonal** of a polygon is any segment that joins two nonconsecutive vertices. Figure 96 shows five-sided polygon *QRSTU.* Segments $\overline{QS}$, $\overline{SU}$, $\overline{UR}$, $\overline{RT}$, and $\overline{QT}$ are the diagonals in this polygon.

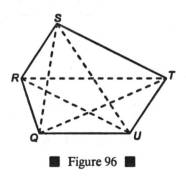

■ Figure 96 ■

- **Number of sides.** Polygons are further classified by how many sides (or angles) they have.

  A **triangle** is a three-sided polygon.
  A **quadrilateral** is a four-sided polygon.
  A **pentagon** is a five-sided polygon.
  A **hexagon** is a six-sided polygon.
  A **septagon** or **heptagon** is a seven-sided polygon.
  An **octagon** is an eight-sided polygon.
  A **nonagon** is a nine-sided polygon.
  A **decagon** is a ten-sided polygon.

It was shown earlier that an equilateral triangle is automatically equiangular and that an equiangular triangle is automatically equilateral. This is not true for polygons in general. Figure 97 shows examples of quadrilaterals that are equiangular but not equilateral, equilateral but not equiangular, and equiangular and equilateral.

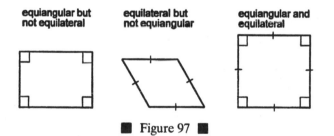

equiangular but not equilateral   equilateral but not equiangular   equiangular and equilateral

■ Figure 97 ■

Regular polygon. When a polygon is both equilateral and equiangular, it is referred to as a **regular polygon.** For a polygon to be regular, it must also be convex. Figure 98 shows examples of regular polygons.

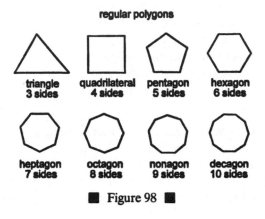

regular polygons

triangle
3 sides

quadrilateral
4 sides

pentagon
5 sides

hexagon
6 sides

heptagon
7 sides

octagon
8 sides

nonagon
9 sides

decagon
10 sides

■ Figure 98 ■

Angle Sums of Polygons

Interior angle sum. When you begin with a polygon with four or more sides and draw all the diagonals possible from one vertex, the polygon then is divided into several nonoverlapping triangles. Figure 99 illustrates this division using a seven-sided polygon. The **interior angle sum** of this polygon can now be found by multiplying the number of triangles by 180°. Upon investigating, it is found that the number of triangles is always two less than the number of sides. This fact is stated as a theorem.

Theorem 39: If a convex polygon has n sides, then its interior angle sum is given by the following equation: $S = (n - 2) \times 180°$.

seven sides

(7 − 2) nonoverlapping triangles = 5 nonoverlapping
triangles

interior angle sum = 5 ×
180°

interior angle sum = 900°

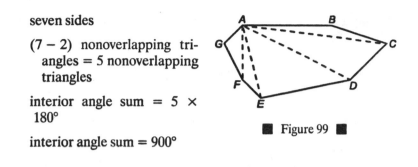

■ Figure 99 ■

Exterior angle of a polygon. An exterior angle of a polygon is formed by extending only one of its sides. The angle adjacent to an interior angle is the exterior angle. Figure 100 might suggest the following theorem.

Theorem 40: If a polygon is convex, then the sum of the degree measures of the exterior angles, one at each vertex, is 360°.

$$m \angle 1 + m \angle 2 + m \angle 3 + m \angle 4 + m \angle 5 + m \angle 6 = 360°$$

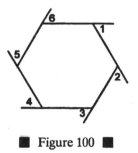

■ Figure 100 ■

Example 1: Find the interior angle sum of a decagon.

A decagon has 10 sides. $S = (10 − 2) \times 180°$

$$S = 1440°$$

Example 2: Find the exterior angle sums, one exterior angle at each vertex, of a nonagon.

The sum of the exterior angles of any polygon is 360°.

Example 3: Find the measure of each interior angle of a regular hexagon.

Method 1: Since the polygon is regular, all interior angles are equal, so you only need to find the interior angle sum and divide by the number of angles.

$$S = (6 - 2) \times 180°$$

$$S = 720$$

There are six angles, so

$$720 \div 6 = 120°$$

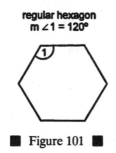

regular hexagon
$m \angle 1 = 120°$

■ Figure 101 ■

Each interior angle of a regular hexagon has a measure of 120°.

Method 2: Since the polygon is regular and all its interior angles are equal, then all its exterior angles are also equal. Look at figure 100. This means that

$$m \angle 1 = m \angle 2 = m \angle 3 = m \angle 4 = m \angle 5 = m \angle 6$$

Since the sum of these angles will always be 360°, then each exterior angle would be 60° (360° ÷ 6 = 60°). If each exterior angle is 60°, then each interior angle is 120° (180° − 60° = 120°).

Special Quadrilaterals

Trapezoid. A **trapezoid** is a quadrilateral with only one pair of opposite sides parallel. The parallel sides are called **bases**, and the nonparallel sides are called **legs**. A segment that joins the midpoints of the legs is called a **median of the trapezoid**. Any segment that is perpendicular to both bases is called the **height** (or **altitude**) **of the trapezoid** (Figure 102).

$\overline{AB}$ and $\overline{CD}$ are bases.

$\overline{XY}$ is a height.

$\overline{MN}$ is a median.

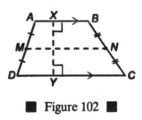

■ Figure 102 ■

Parallelogram. A **parallelogram** is any quadrilateral with both pairs of opposite sides parallel. Each pair of parallel sides are called **bases of the parallelogram.** Any perpendicular segment between a pair of bases is called a **height of the parallelogram.** The symbol ▱ is used for the word *parallelogram.* Figure 103 shows that a parallelogram has two sets of bases and that, with each set of bases, there is an associated height.

In ▱$ABCD$,

$\overline{XY}$ is a height to bases $\overline{AB}$ and $\overline{CD}$.

$\overline{JK}$ is a height to bases $\overline{AD}$ and $\overline{BC}$.

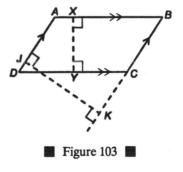

■ Figure 103 ■

The following are theorems regarding parallelograms.

Theorem 41: A diagonal of a parallelogram divides it into two congruent triangles.

In □*ABCD* with diagonal $\overline{BD}$,

according to *Theorem 41,*

$\triangle ABD \cong \triangle CDB$.

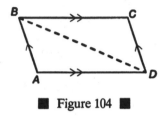

■ Figure 104 ■

Theorem 42: Opposite sides of a parallelogram are equal.

Theorem 43: Opposite angles of a parallelogram are equal.

Theorem 44: Consecutive angles of a parallelogram are supplementary.

In □*ABCD,*

by *Theorem 42, AB = DC* and *AD = BC*

by *Theorem 43, m ∠A = m ∠C* and
$m ∠B = m ∠D$

by *Theorem 44,*

$\left. \begin{array}{l} ∠A \text{ and } ∠B, \\ ∠B \text{ and } ∠C, \\ ∠C \text{ and } ∠D \\ ∠A \text{ and } ∠D, \end{array} \right\}$ are supplementary

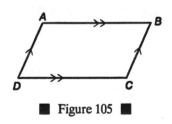

■ Figure 105 ■

Theorem 45: The diagonals of a parallelogram bisect each other.

In □*ABCD*,

by *Theorem 45*, *AE = EC* and
BE = ED

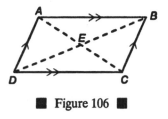

■ Figure 106 ■

Proving That Figures Are Parallelograms

Many times you will be asked to prove that a figure is a parallelogram. The following theorems are tests that determine that a quadrilateral is a parallelogram.

Theorem 46: If both pairs of opposite sides of a quadrilateral are equal, then it is a parallelogram.

Theorem 47: If both pairs of opposite angles of a quadrilateral are equal, then it is a parallelogram.

Theorem 48: If all pairs of consecutive angles of a quadrilateral are supplementary, then it is a parallelogram.

Theorem 49: If one pair of opposite sides of a quadrilateral are both equal and parallel, then it is a parallelogram.

Theorem 50: If the diagonals of a quadrilateral bisect each other, then it is a parallelogram.

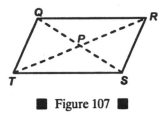

■ Figure 107 ■

Quadrilateral *QRST* in Figure 107 is a parallelogram if

(1) $QR = ST$ and $QT = RS$, by *Theorem 46*

(2) $m \angle Q = m \angle S$ and $m \angle T = m \angle R$, by *Theorem 47*

(3) $\angle Q$ and $\angle R$, $\angle R$ and $\angle S$, $\angle S$ and $\angle T$, and $\angle Q$ and $\angle T$ are all supplementary pairs, by *Theorem 48*

(4) $QR = ST$ and $\overline{QR} \parallel \overline{ST}$ or $QT = RS$ and $\overline{QT} \parallel \overline{RS}$, by *Theorem 49*

(5) $QP = PS$ and $RP = PT$, by *Theorem 50*

Properties of Special Parallelograms

Rectangle. A rectangle is a quadrilateral with all right angles. It is easily shown that it must also be a parallelogram, with all of its properties. But a rectangle has an additional property.

Theorem 51: The diagonals of a rectangle are equal.

In rectangle *ABCD*,

$AC = BD$, by *Theorem 51*

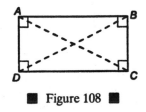

■ Figure 108 ■

Rhombus. A **rhombus** is a quadrilateral with all equal sides. It is also a parallelogram with all its properties. But a rhombus also has additional properties.

Theorem 52: The diagonals of a rhombus bisect opposite angles.

Theorem 53: The diagonals of a rhombus are perpendicular to one another.

In rhombus *CAND*,

by *Theorem 52,*

$\overline{CN}$ bisects ∠*DCA* and ∠*DNA*, also

$\overline{AD}$ bisects ∠*CAN* and ∠*CDN*

and by *Theorem 53,*

$\overline{CN} \perp \overline{AD}$

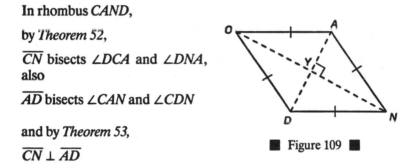

■ Figure 109 ■

Square. A **square** is a quadrilateral with all right angles and all equal sides. A square is also a parallelogram, a rectangle, and a rhombus and has all the properties of all these special quadrilaterals. Figure 110 shows a square.

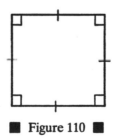

■ Figure 110 ■

The relationships of these quadrilaterals to one another is summarized in Figure 111.

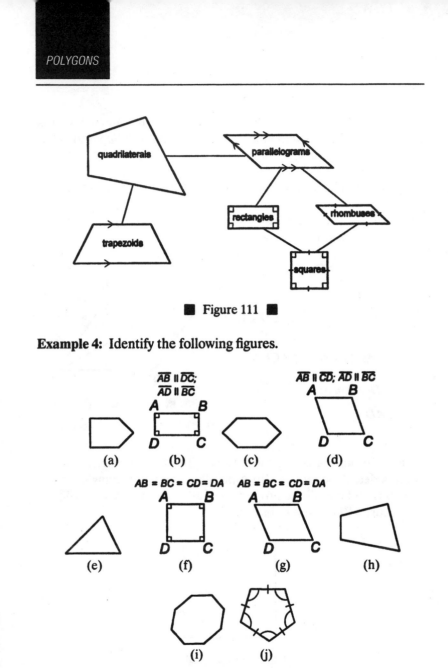

■ Figure 111 ■

Example 4: Identify the following figures.

■ Figure 112 ■

(a) pentagon, (b) rectangle, (c) hexagon, (d) parallelogram, (e) triangle, (f) square, (g) rhombus, (h) quadrilateral, (i) octagon, (j) regular pentagon

Example 5: In Figure 113, find $m \angle A$, $m \angle C$, $m \angle D$, CD, and AD.

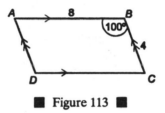

■ Figure 113 ■

$m \angle A = 80°$⎤ Consecutive angles of a parallelogram are
$m \angle C = 80°$⎦ supplementary.

$m \angle D = 100°$} Opposite angles of a parallelogram are equal.

$CD = 8$⎤
$AD = 4$⎦ Opposite sides of a parallelogram are equal.

Example 6: In Figure 114, find *TR*, *QP*, *PS*, *TP*, and *PR*.

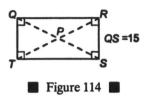

■ Figure 114 ■

TR = 15} Diagonals of a rectangle are equal.

$$
\left.\begin{array}{l}
QP = 7.5 \\
PS = 7.5 \\
TP = 7.5 \\
PR = 7.5
\end{array}\right\}
$$ Diagonals of a rectangle bisect each other.

Example 7: In Figure 115, find *m ∠MOE*, *m ∠NOE*, and *m ∠MYO*.

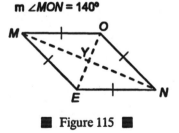

■ Figure 115 ■

$\left.\begin{array}{l} m\ \angle MOE = 70° \\ m\ \angle NOE = 70° \end{array}\right\}$ Diagonals of a rhombus bisect opposite angles.

m ∠MYO = 90°} Diagonals of a rhombus are perpendicular.

Properties of Trapezoids

Isosceles trapezoid. Recall that a trapezoid is a quadrilateral with only one pair of opposite sides parallel and that the parallel sides are called bases and the nonparallel sides are called legs. If the legs of a trapezoid are equal, it is called an **isosceles trapezoid**. Figure 116 is an isosceles trapezoid.

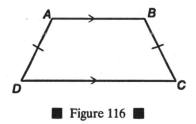

■ Figure 116 ■

Base angles. A pair of angles sharing the same base are called **base angles** of the trapezoid. In Figure 116, ∠A and ∠B or ∠C and ∠D are base angles of trapezoid *ABCD*. Two special properties of an isosceles trapezoid can be proven.

Theorem 53: Base angles of an isosceles trapezoid are equal.

Theorem 54: Diagonals of an isosceles trapezoid are equal.

In isosceles trapezoid *ABCD* with bases $\overline{AB}$ and $\overline{CD}$,

by *Theorem 53,*

$m \angle DAB = m \angle CBA$
$m \angle ADC = m \angle BCD$

and by *Theorem 54,*

$AC = BD$

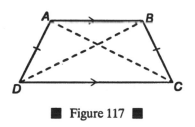

■ Figure 117 ■

Recall that the median of a trapezoid is a segment that joins the midpoints of the nonparallel sides.

Theorem 55: The median of any trapezoid has two properties: (1) It is parallel to both bases. (2) Its length equals half the sum of the base lengths.

In trapezoid *ABCD* with bases $\overline{AB}$ and $\overline{CD}$, *E* the midpoint of $\overline{AD}$, and *F* the midpoint of $\overline{BC}$,

by *Theorem 55,*

$\overline{EF} \parallel \overline{AB}$

$\overline{EF} \parallel \overline{CD}$

$EF = \frac{1}{2}(AB + CD)$

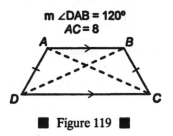

■ Figure 118 ■

Example 8: In Figure 119, find $m \angle ABC$ and find *BD*.

m ∠DAB = 120°
AC = 8

■ Figure 119 ■

$m \angle ABC = 120°$} Base angles of an isosceles trapezoid are equal.

$BD = 8$} Diagonals of an isosceles trapezoid are equal.

Example 9: In Figure 120, find *TU*.

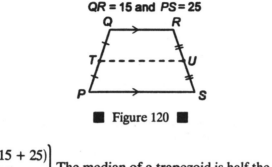

QR = 15 and PS = 25

■ Figure 120 ■

$$TU = \tfrac{1}{2}(15 + 25)$$
$$= \tfrac{1}{2}(40)$$
$$= 20$$

The median of a trapezoid is half the sum of the lengths of the bases.

The Midpoint Theorem

Figure 121 shows $\triangle ABC$ with D and E as midpoints of sides $\overline{AC}$ and $\overline{AB}$ respectively. If you look at this triangle as though it were a trapezoid with one base of $\overline{BC}$ and the other base so small that its length is virtually zero, you could apply the "median" theorem of trapezoids, *Theorem 55*.

Theorem 56: (Midpoint Theorem) The segment joining the midpoints of two sides of a triangle is parallel to the third side and half as long as the third side.

In Figure 121,

by *Theorem 56*,

$\overline{DE} \parallel \overline{BC}$

$DE = \tfrac{1}{2}BC$

■ Figure 121 ■

Example 10: In Figure 122, find *HJ*.

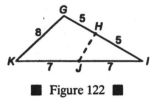

■ Figure 122 ■

$HJ = \frac{1}{2}(8)$ Since *H* and *J* are midpoints of two sides of a triangle,
$HJ = 4$ $HJ = \frac{1}{2}GK$.

Perimeter refers to the distance around a figure. If the figure is a circle, that distance is referred to as a **circumference.** Distance around is always measured in linear units such as inches, feet, or centimeters. **Area** refers to the space inside a plane (flat) figure. Area is always measured in square units such as square inches (in^2), square feet (ft^2) or square centimeters (cm^2).

Squares and Rectangles

For all polygons, perimeter is found by adding together the lengths of all the sides. In this section, P will be used to stand for *perimeter,* and A will be used to stand for *area.*

Perimeter of a square and perimeter of a rectangle. Perimeter formulas for squares and rectangles are easily seen from Figures 123(a) and 123(b).

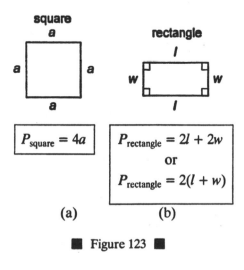

$$P_{square} = 4a$$

$$P_{rectangle} = 2l + 2w$$

or

$$P_{rectangle} = 2(l + w)$$

(a)   (b)

■ Figure 123 ■

Area of a square and area of a rectangle. Area formulas for squares and rectangles are formed by simply multiplying any pair of consecutive sides together. Refer to Figures 123(a) and 123(b).

$$A_{\text{square}} = a^2 \qquad A_{\text{rectangle}} = lw$$

Example 1: Find the perimeter and area of Figure 124.

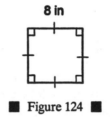

8 in

■ Figure 124 ■

This is a square.

$$P_{\text{square}} = 4a \qquad A_{\text{square}} = a^2$$
$$= 4(8 \text{ in}) \qquad = (8 \text{ in})^2$$
$$= 32 \text{ in} \qquad = 64 \text{ in}^2$$

Example 2: Find the perimeter and area of Figure 125.

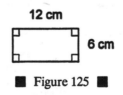

12 cm

6 cm

■ Figure 125 ■

This is a rectangle.

$P_{rectangle} = 2(l + w)$ $A_{rectangle} = lw$

$= 2(12\text{ cm} + 6\text{ cm})$ $= (12\text{ cm})(6\text{ cm})$

$= 36\text{ cm}$ $= 72\text{ cm}^2$

Example 3: If the perimeter of a square is 36 ft, find its area.

$P_{square} = 4a$ $A_{square} = a^2$

$36\text{ ft} = 4a$ $= (9\text{ ft})^2$

$9\text{ ft} = a$ $= 81\text{ ft}^2$

The area of the square would be 81 square feet.

Example 4: If a rectangle with length 9 in has an area of 36 in², find its perimeter.

$A_{rectangle} = lw$ $P_{rectangle} = 2(l + w)$

$36\text{ in}^2 = (9\text{ in})(w)$ $= 2(9\text{ in} + 4\text{ in})$

$4\text{ in} = w$ $= 26\text{ in}$

The perimeter of the rectangle would be 26 inches.

Parallelograms

In this parallelogram, Figure 126, h is a height, since it is perpendicular to a pair of opposite sides called bases. One of the bases has been labeled b. The remaining sides are each labeled a.

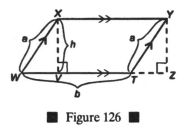

■ Figure 126 ■

Perimeter of a parallelogram. The following formula is now apparent.

$$P_{\text{parallelogram}} = 2(a + b) \quad \text{or} \quad P_{\text{parallelogram}} = 2a + 2b$$

Area of a parallelogram. In Figure 126, also notice that $\triangle WXV \cong \triangle TYZ$, which means that they also have equal areas. This makes the area of $\square WXYT$ the same as the area of $\square XYZV$. But $A_{\text{rectangle } XYZV} = bh$, so $A_{\text{parallelogram } WXYZ} = bh$. That is, the area of a parallelogram is the product of any base with its respective height.

$$A_{\text{parallelogram}} = bh$$

Example 5: Find the perimeter and area of Figure 127.

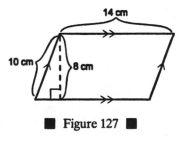

■ Figure 127 ■

The figure is a parallelogram

$P_{\text{parallelogram}} = 2(a + b)$

$\quad = 2(10 \text{ cm} + 14 \text{ cm})$

$\quad = 48 \text{ cm}$

$A_{\text{parallelogram}} = bh$

$\quad = (14 \text{ cm})(8 \text{ cm})$

$\quad = 112 \text{ cm}^2$

Triangles

Look at $\triangle ABD$ in Figure 128. If a line is drawn through B parallel to $\overline{AD}(\overline{BC})$ and another line is drawn through D parallel to $\overline{AB}$ ($\overline{DC}$), then you will have formed a parallelogram. $\overline{BD}$ is now a diagonal in this parallelogram. Since a diagonal divides a parallelogram into two congruent triangles, the area of $\triangle ABD$ is exactly half the area of $\square ABCD$.

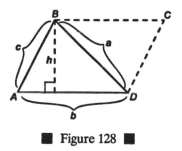

■ Figure 128 ■

Area of a triangle. Since $A_{\text{parallelogram}} = bh$, then

$$A_{\text{triangle}} = \tfrac{1}{2}bh$$

Perimeter of a triangle. In $\triangle ABD$ (Figure 128), the perimeter is found simply by adding the lengths of the three sides.

$$\boxed{P_{\text{triangle}} = a + b + c}$$

Example 6: Find the perimeter and area for the triangles in Figures 129(a), 129(b), and 129(c).

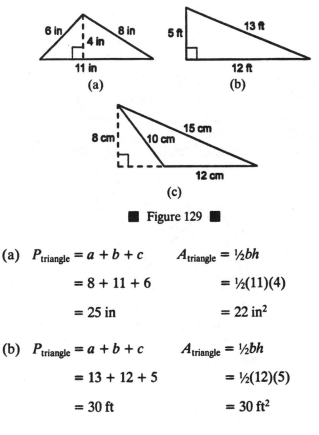

■ Figure 129 ■

(a) $P_{\text{triangle}} = a + b + c$ $A_{\text{triangle}} = \frac{1}{2}bh$

       $= 8 + 11 + 6$ $= \frac{1}{2}(11)(4)$

       $= 25$ in $= 22$ in^2

(b) $P_{\text{triangle}} = a + b + c$ $A_{\text{triangle}} = \frac{1}{2}bh$

          $= 13 + 12 + 5$ $= \frac{1}{2}(12)(5)$

          $= 30$ ft $= 30$ ft^2

(c) $P_{\text{triangle}} = a + b + c$ $A_{\text{triangle}} = \frac{1}{2}bh$

          $= 15 + 12 + 10$ $= \frac{1}{2}(12)(8)$

          $= 37$ cm $= 48$ cm^2

Example 7: If the area of a triangle is 64 cm^2 and it has a height of 16 cm, find the length of its base.

$$A_{\text{triangle}} = \tfrac{1}{2}bh$$
$$64 \text{ cm}^2 = \tfrac{1}{2}(b)(16 \text{ cm})$$

Multiply both sides by 2.

$$128 \text{ cm}^2 = (b)(16 \text{ cm})$$
$$8 \text{ cm} = b$$

The triangle will have a base of 8 centimeters.

Trapezoids

Perimeter of a trapezoid. In Figure 130, trapezoid $QRSV$ is labeled so that b_1 and b_2 are the bases (h is the height to these bases) and a and c are the legs. The perimeter is simply the sum of these lengths.

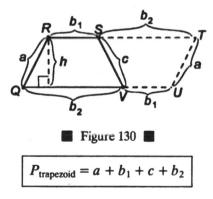

■ Figure 130 ■

$$P_{\text{trapezoid}} = a + b_1 + c + b_2$$

Area of a trapezoid. Referring to Figure 130, an identical, but upside-down trapezoid is drawn adjacent to trapezoid *QRSV*, trapezoid *TUVS*. It can now be shown that the figure *QRTU* is a parallelogram, and its area can now be found.

$$A_{\text{parallelogram } QRTU} = (\text{base})(\text{height})$$

$$= (b_1 + b_2)h$$

Since trapezoid *QRSV* is exactly half of this parallelogram, the following formula gives the area of a trapezoid.

$$\boxed{A_{\text{trapezoid}} = \tfrac{1}{2}(b_1 + b_2)h}$$

Example 8: Find the perimeter and area of Figure 131.

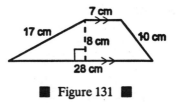

■ Figure 131 ■

The figure is a trapezoid.

$$P_{\text{trapezoid}} = a + b_1 + c + b_2 \qquad A_{\text{trapezoid}} = \tfrac{1}{2}(b_1 + b_2)h$$

$$= 17 + 7 + 10 + 28 \qquad\qquad = \tfrac{1}{2}(7 + 28)(8)$$

$$= 62 \text{ cm} \qquad\qquad\qquad = 140 \text{ cm}^2$$

Regular Polygons

Perimeter of a regular polygon. Since a regular polygon is equilateral, to find its perimeter you need to know only the length of one of its sides and multiply that by the number of sides. Using *n*-gon to represent a polygon with *n* sides, and *s* as the length of each side, produces the following formula.

$$P_{\text{regular } n\text{-gon}} = ns$$

Center of the regular polygon. Figure 132 is a regular polygon.

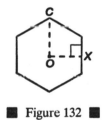

■ Figure 132 ■

In a regular polygon, there is one point in its interior that is equidistant from its vertices. This point is called the **center of the regular polygon.** In Figure 132, *O* is the center of the regular polygon.

Radius of a regular polygon. A **radius of a regular polygon** is a segment that goes from the center to any vertex of the regular polygon.

Apothem of a regular polygon. An **apothem of a regular polygon** is any segment that goes from the center and is perpendicular to one of the polygon's sides. In Figure 132, $\overline{OC}$ is a radius and $\overline{OX}$ is an apothem.

Area of a regular polygon. If p represents the perimeter of the regular polygon and a represents the length of its apothem, the following formula can eventually be shown to represent its area.

$$A_{\text{regular } n\text{-gon}} = \tfrac{1}{2}ap$$

Example 9: Find the perimeter and area of the regular pentagon in Figure 133.

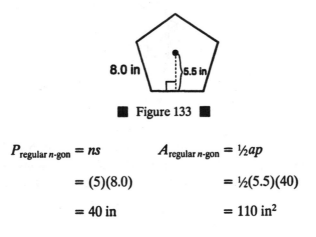

8.0 in   5.5 in

■ Figure 133 ■

$P_{\text{regular } n\text{-gon}} = ns$ $\qquad$ $A_{\text{regular } n\text{-gon}} = \tfrac{1}{2}ap$

$\qquad = (5)(8.0)$ $\qquad\qquad = \tfrac{1}{2}(5.5)(40)$

$\qquad = 40 \text{ in}$ $\qquad\qquad = 110 \text{ in}^2$

Circles

A **circle** is a plane figure with all points the same distance from one point. That one point is called the **center of the circle.** Any segment that goes from the center to a point on the circle is called a **radius of the circle.** A **diameter** is any segment which passes through the center and has its endpoints on the circle. It is now evident that a diameter is twice as long as a radius. In Figure 134, O is the center, $\overline{OB}$, $\overline{OC}$, and $\overline{OA}$ are each a radius, and $\overline{AC}$ is a diameter.

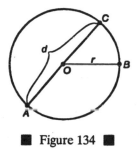

■ Figure 134 ■

Circumference of a circle. Greeks in ancient times found that the circumference of any circle divided by its diameter always yields the same value. The Greek letter π (pi) is now used to represent that value. In fractional or decimal form, the commonly used approximations of π are $\pi \approx 3.14$ or $\pi \approx 22/7$. The Greeks found the formula $C_{circle}/d = \pi$, which is rewritten in the following form.

$$C_{circle} = \pi d \quad \text{or} \quad C_{circle} = 2\pi r$$

If you briefly look at a circle as a regular polygon with infinitely many sides, you see that the apothem and radius become the same length (Figure 135).

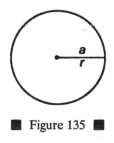

■ Figure 135 ■

Area of a circle. Looking at the area formula for a regular polygon and making the appropriate changes with regard to the circle.

$$A_{\text{regular } n\text{-gon}} = \tfrac{1}{2}(a)(p)$$
$$A_{\text{circle}} = \tfrac{1}{2}(r)(2\pi r)$$
$$= \tfrac{1}{2}(2\pi r)(r)$$
$$= \pi r^2$$

That is, the formula for the area of a circle now becomes the following.

$$\boxed{A_{\text{circle}} = \pi r^2}$$

Example 10: Find the circumference and area for the circle in Figure 136. Use 3.14 as an approximation for π.

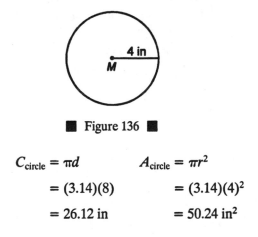

■ Figure 136 ■

$$C_{\text{circle}} = \pi d \qquad A_{\text{circle}} = \pi r^2$$
$$= (3.14)(8) \qquad = (3.14)(4)^2$$
$$= 26.12 \text{ in} \qquad = 50.24 \text{ in}^2$$

Example 11: If the area of a circle is 81π ft^2, find its circumference.

$$A_{\text{circle}} = \pi r^2 \qquad C_{\text{circle}} = \pi d$$
$$81\pi \text{ ft}^2 = \pi r^2 \qquad\qquad = \pi(18)$$
$$9 \text{ ft} = r \qquad\qquad = 18\pi \text{ ft}$$
$$d = 18 \text{ ft}$$

So the circumference is 18 ft.

Summary of Perimeter/Circumference and Area Formulas

Figure	Name	Perimeter/ Circumference	Area
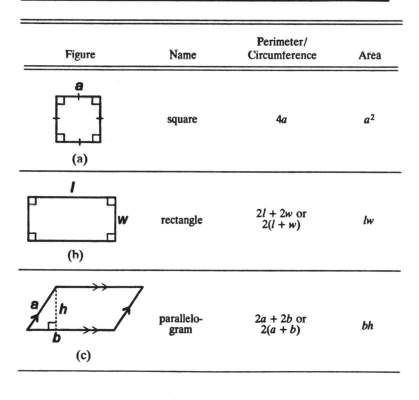 square (a)	square	$4a$	a^2
rectangle (b)	rectangle	$2l + 2w$ or $2(l + w)$	lw
parallelogram (c)	parallelo-gram	$2a + 2b$ or $2(a + b)$	bh

Figure	Name	Perimeter/Circumference	Area
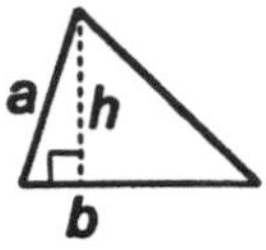 (d)	triangle	$a + b + c$	$\frac{1}{2}bh$
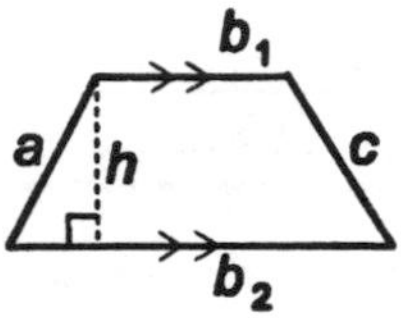 (e)	trapezoid	$a + b_1 + c + b_2$	$\frac{1}{2}(b_1 + b_2)h$
 (f)	regular polygon	ns n = number of sides	$\frac{1}{2}ap$ p = perimeter a = apothem
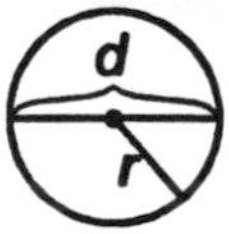 (g)	circle	πd or $2\pi r$	πr^2

■ Figure 137 ■

Ratio and Proportion

Ratio. A **ratio** is a comparison of two numbers usually expressed in simplest fraction form. Another form uses a colon. The colon form is most frequently used when comparing three or more numbers to each other.

Ratio	Written Form
3 to 4	3/4 or 3 : 4
a to b, $b \neq 0$	a/b or $a : b$
1 to 3 to 5	1 : 3 : 5

Example 1: A classroom has 25 boys and 15 girls. What is the ratio of boys to girls?

$$\text{boys to girls} = 25 \text{ to } 15 \quad \text{or} \quad 25 : 15$$

Reduce $\qquad\qquad\qquad = \;\; 5 \text{ to } 3 \quad \text{or} \quad 5 : 3$

The ratio of boys to girls is 5 to 3, or 5/3, or 5 : 3.

Example 2: The ratio of two supplementary angles is 2 to 3. Find the measure of each angle.

Let

$\left.\begin{array}{l}\text{measure of smaller angle} = 2x \\ \text{measure of larger angle} = 3x\end{array}\right\}$ $2x$ to $3x$ will reduce to 2 to 3.

$2x + 3x = 180°\}$ The sum of supplementary angles is 180°.

$$5x = 180°$$
$$x = 36°$$

Then $\quad 2x = 2(36°) \quad$ and $\quad 3x = 3(36°)$

$\qquad\qquad 2x = 72° \qquad$ and $\quad 3x = 108°$

The angles have measures of 72° and 108°.

Example 3: A triangle has angle measures of 40°, 50°, and 90°. In simplest form, what is the ratio of these angles to each other?

$$40 : 50 : 90 = 4 : 5 : 9 \quad \text{(10 is a common divisor)}$$

This means that

(1) the ratio of the first to the second is 4 to 5
(2) the ratio of the first to the third is 4 to 9
(3) the ratio of the second to the third is 5 to 9

Example 4: A 50-inch segment is divided into three parts whose lengths have the ratio 2 : 3 : 5. What is the length of the longest part?

Let $\qquad$ measure of shortest piece = $2x$
$\qquad\qquad$ measure of middle piece = $3x$
$\qquad\qquad$ measure of longest piece = $5x$

$$2x + 3x + 5x = 50$$
$$10x = 50$$
$$x = 5$$

$2x = 2(5) \qquad 3x = 3(5) \qquad 5x = 5(5)$

$2x = 10 \qquad\; 3x = 15 \qquad\;\; 5x = 25$

The longest part has a measure of 25 inches.

Proportion. A **proportion** is an equation showing that two ratios are equal.

$$\frac{8}{10} = \frac{4}{5} \qquad 8 : 10 = 4 : 5$$

Means and extremes.

In the above proportion, the values a and d are called **extremes** of the proportion; the values b and c are called the **means** of the proportion.

Properties of Proportions

Property 1: If $a/b = c/d$, then $ad = bc$ (*Means-Extremes Property*, or *Cross-Products Property*).

$$8/10 = 4/5 \text{ is a proportion}$$

Property 1 states $\quad (8)(5) = (10)(4)$

$$40 = 40$$

Example 5: Find a if $a/12 = 3/4$.

By *Property 1*, $(a)(4) = (12)(3)$

$$4a = 36$$
$$a = 9$$

Example 6: Is $3 : 4 = 7 : 8$ a proportion?

No. If this were a proportion, *Property 1* would produce

$$(3)(8) = (4)(7)$$
$$24 = 28 \quad \text{which is } not \text{ true}$$

Property 2: If $a/b = c/d$, then $d/b = c/a$ (*Means or Extremes Switching Property*).

Example 7: $8/10 = 4/5$ is a proportion. *Property 2* says that if you were to switch the 8 and 5 or switch the 4 and 10, then the new statement is still a proportion.

If $8/10 = 4/5$, then $5/10 = 4/8$, or if $8/10 = 4/5$, then $8/4 = 10/5$.

Example 8: If $x/5 = y/4$, find the ratio of x/y.

$$x/5 = y/4$$

Use the switching property of proportions and switch the means positions, the 5 and the y.

$$x/y = 5/4$$

Property 3: If $a/b = c/d$, then $b/a = d/c$ (*Upside-Down Property*).

Example 9: If $9a = 5b$, find the ratio a/b.

Rewrite $9a = 5b$ in its proportion form so that $9a = 5b$ is the result of applying the *Cross-Products Property.*

$$9/5 = b/a \quad \text{or} \quad 9/b = 5/a.$$

Apply *Property 3.*	Apply *Property 2.*
Turn each side upside-down.	Switch the 9 and the a.

So $\quad 5/9 = a/b$, or $a/b = 5/9 \qquad\qquad a/b = 5/9$

Property 4: If $a/b = c/d$, then

$$\left.\begin{array}{c}(a + b)/b = (c + d)/d \\ \text{or} \\ (a - b)/b = (c - d)/d\end{array}\right\} \text{Demoninator Addition/Subtraction Property}$$

Example 10: If $5/8 = x/y$, then $13/8 = ?$

$$5/8 = x/y$$

Apply *Property 4.* $\quad (5 + 8)/8 = (x + y)/y$

$$13/8 = (x + y)/y$$

Example 11: In Figure 138, $AB/BC = 5/8$. Find AC/BC.

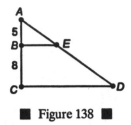

■ Figure 138 ■

Recall that $AB + BC = AC$ (*Segment Addition Postulate*).

If $$\frac{AB}{BC} = \frac{5}{8}$$

then $$\frac{AB + BC}{BC} = \frac{5 + 8}{8} \quad (Property\ 4)$$

So $$\frac{AC}{BC} = \frac{13}{8}$$

Example 12: A map is scaled so that 3 cm on the map is equal to 5 actual miles. If two cities on the map are 10 cm apart, what is the actual distance the cities are apart?

Let x = the actual distance.

$$\frac{\text{map}}{\text{actual miles}} = \frac{3}{5} = \frac{10}{x}$$

Apply the *Cross-Products Property*.

$$3x = 50$$

$$x = 16\tfrac{2}{3}$$

The cities are $16\tfrac{2}{3}$ miles apart.

Similar Polygons

Two polygons with the same shape are called **similar polygons.** The symbol for "is similar to" is ~. Notice that it is a portion of the "is congruent to" symbol, ≅. When two polygons are similar, two facts *both* must be true.

(1)   Corresponding angles are equal.
(2)   The ratios of each pair of corresponding sides are equal.

In Figure 139, quadrilateral *ABCD* ~ quadrilateral *EFGH*.

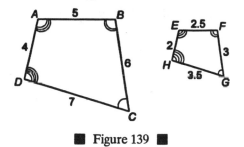

■ Figure 139 ■

This means: $m \angle A = m \angle E$, $m \angle B = m \angle F$, $m \angle C = m \angle G$, $m \angle D = m \angle H$, and

$$\frac{AB}{EF} = \frac{BC}{FG} = \frac{CD}{GH} = \frac{AD}{EH}$$

It is possible for a polygon to have one of the above facts true without having the other fact true. The following two examples show how that is possible.

In Figure 140, quadrilateral *QRST* is not similar to quadrilateral *WXYZ*.

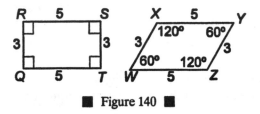

■ Figure 140 ■

Even though the ratios of corresponding sides are equal, corresponding angles are not equal (90° ≠ 120°, 90° ≠ 60°).

In Figure 141, quadrilateral *FGHI* is not similar to quadrilateral *JKLM*.

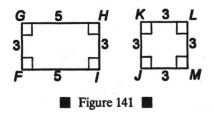

■ Figure 141 ■

Even though corresponding angles are equal, the ratios of each pair of corresponding sides are not equal (3/3 ≠ 5/3).

Example 13: In Figure 142, quadrilateral *ABCD* ~ quadrilateral *EFGH*. (a) Find *m* ∠*E*. (b) Find *x*.

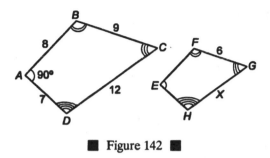

■ Figure 142 ■

(a) *m* ∠*E* = 90° (∠*E* and ∠*A* are corresponding angles of similar polygons, and corresponding angles of similar polygons are equal.)

(b) 9/6 = 12/*x*. (If two polygons are similar, then the ratios of each pair of corresponding sides are equal.)

$$9x = 108 \; \textit{(Cross-Products Property)}$$

$$x = 12$$

Similar Triangles

In general, to prove that two polygons are similar, you must show that *all* pairs of corresponding angles are equal and that *all* ratios of each pair of corresponding sides are equal. In triangles, though, this is not necessary.

Postulate 17: (*AA Similarity Postulate*) If two angles of one triangle are equal to two angles of another triangle, then the triangles are similar.

Example 14: Use Figure 143 to show that the triangles are similar.

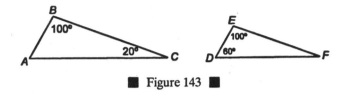

■ Figure 143 ■

$m \angle B = m \angle E$

In $\triangle ABC$, $m \angle A + m \angle B + m \angle C = 180°$

$$m \angle A + 100° + 20° = 180°$$

$$m \angle A = 60°$$

But in $\triangle DEF$, $m \angle D = 60°$

So $m \angle A = m \angle D$

By *Postulate 17,* the *AA Similarity Postulate,* $\triangle ABC \sim \triangle DEF$. Additionally, since the triangles are now similar,

$$m \angle C = m \angle F$$

and $\dfrac{AB}{DE} = \dfrac{BC}{EF} = \dfrac{AC}{DF}$

Example 15: Use Figure 144 to show that $\triangle QRS \sim \triangle UTS$.

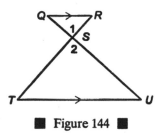

■ Figure 144 ■

$m \angle 1 = m \angle 2$} Vertical angles are equal.

$m \angle R = m \angle T$
or
$m \angle Q = m \angle U$ } If two parallel lines are cut by a transversal, then the alternate interior angles are equal.

So by the *AA Similarity Postulate*, $\triangle QRS \sim \triangle UTS$.

Example 16: Use Figure 145 to show that $\triangle MNO \sim \triangle PQR$.

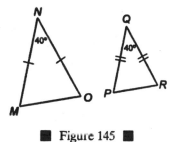

■ Figure 145 ■

In $\triangle MNO, MN = NO$, and in $\triangle PQR, PQ = QR$.

$m \angle M = m \angle O$ and $m \angle P = m \angle R$

(If two sides of a triangle are equal, the angles opposite these sides have equal measures.)

In △MNO,   In △PQR,

$m \angle M + m \angle N + m \angle O = 180°$ $m \angle P + m \angle Q + m \angle R = 180°$

Since $m \angle M = m \angle O$ and $m \angle P = m \angle R$,

$$2m \angle M + 40° = 180° \qquad 2m \angle P + 40° = 180°$$
$$2m \angle M = 140° \qquad 2m \angle P = 140°$$
$$m \angle M = 70° \qquad m \angle P = 70°$$

So $m \angle M = m \angle P$, and $m \angle O = m \angle R$. △MNO ~ △PQR (*AA Similarity Postulate*).

Example 17: Use Figure 146 to show that △ABC ~ △DEF.

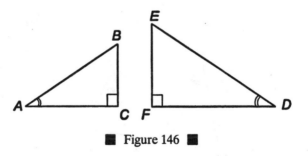

■ Figure 146 ■

$m \angle C = m \angle F$ (All right angles are equal.)

$m \angle A = m \angle D$ (They are indicated as equal in the figure.)

△ABC ~ △DEF   (*AA Similarity Postulate*)

Proportional Parts of Triangles

Consider Figure 147 of $\triangle ABC$ with line l parallel to $\overline{AC}$ and intersecting the other two sides at D and E.

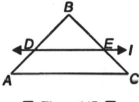

■ Figure 147 ■

You can eventually prove that $\triangle ABC \sim \triangle DBE$ using the *AA Similarity Postulate*. Since the ratios of corresponding sides of similar polygons are equal, you can show that

$$\frac{AB}{BD} = \frac{BC}{BE}$$

Now use *Property 4*, the *Denominator Subtraction Property*.

$$\frac{AB - DB}{BD} = \frac{BC - BE}{BE}$$

But $AB - DB = AD$, and $BC - BE = CE$ (*Segment Addition Postulate*). With this replacement, you get the following proportion.

$$\frac{AD}{BD} = \frac{CE}{BE}$$

This leads to the following theorem.

Theorem 57: (*Side-Splitter Theorem*) If a line is parallel to one side of a triangle and intersects the other two sides, it divides those sides proportionally.

Example 18: Use Figure 148 to find *x*.

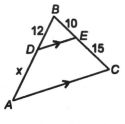

■ Figure 148 ■

Since $\overline{DE} \parallel \overline{AC}$ in $\triangle ABC$, by *Theorem 57,* you get

$$\frac{x}{12} = \frac{15}{10}$$

$$10x = 180 \quad (\textit{Cross-Products Property})$$

$$x = 18$$

Example 19: Use Figure 149 to find *x*.

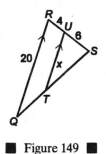

■ Figure 149 ■

Notice that $\overline{TU}$, (*x*), is *not* one of the segments on either side that $\overline{TU}$ intersects. This means that you *cannot* apply *Theorem 57* to this situation. So what can you do? Recall that with $\overline{TU} \parallel \overline{QR}$, you can show that $\triangle QRS \sim \triangle TUS$. Since the ratios of corresponding sides of similar triangles are equal, you get the following proportion.

$$\frac{QR}{TU} = \frac{RS}{US} = \frac{QS}{TS}$$

$$\frac{20}{x} = \frac{10}{6} = \frac{QS}{TS}$$

$$10x = 120 \quad (\text{Cross-Products Property})$$

$$x = 12$$

Another theorem involving parts of a triangle is more complicated to prove but is presented here so you can use it to solve problems related to it.

Theorem 58: (*Angle Bisector Theorem*) If a ray bisects an angle of a triangle, then it divides the opposite side into segments which are proportional to the sides that formed the angle.

In Figure 150, $\overline{BD}$ bisects $\angle ABC$ in $\triangle ABC$. By Theorem 58,

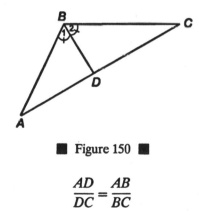

■ Figure 150 ■

$$\frac{AD}{DC} = \frac{AB}{BC}$$

Example 20: Use Figure 151 to find x.

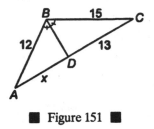

■ Figure 151 ■

Since $\overline{BD}$ bisects $\angle ABC$ in $\triangle ABC$, you apply *Theorem 58.*

$$\frac{x}{13} = \frac{12}{15}$$

$$15x = 156 \quad (\textit{Cross-Products Property})$$

$$x = 10\tfrac{2}{5} \text{ or } 10.4$$

Proportional Parts of Similar Triangles

Theorem 59: If two triangles are similar, then the ratio of any two corresponding segments equals the ratio of any two corresponding sides.

In Figure 152, suppose $\triangle QRS \sim \triangle TUV$.

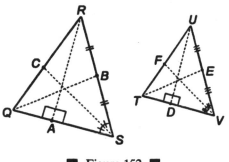

■ Figure 152 ■

Then,
$$\frac{QR}{TU} = \frac{RS}{UV} = \frac{QS}{TV}$$

Then according to *Theorem 59,*

$$\frac{\text{length of altitude } \overline{RA}}{\text{length of altitude } \overline{UD}} = \frac{QR}{TU}$$

$$\frac{\text{length of median } \overline{QB}}{\text{length of median } \overline{TE}} = \frac{QR}{TU}$$

$$\frac{\text{length of } \angle \text{ bisector } \overline{CS}}{\text{length of } \angle \text{ bisector } \overline{FV}} = \frac{QR}{TU}$$

Example 21: Use Figure 153 and the fact that $\triangle ABC \sim \triangle GHI$ to find x.

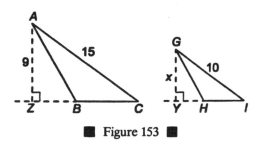

■ Figure 153 ■

$$\frac{AZ}{GY} = \frac{AC}{GI} \quad (\textit{Theorem 59})$$

$$\frac{9}{x} = \frac{15}{10}$$

$$15x = 90 \quad (\textit{Cross-Products Property})$$

$$x = 6$$

Perimeters and Areas of Similar Triangles

When two triangles are similar, the reduced ratio of any two corresponding sides is called the **scale factor** of the similar triangles. In Figure 154, $\triangle ABC \sim \triangle DEF$.

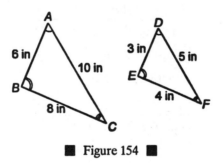

■ Figure 154 ■

The ratios of corresponding sides are 6/3, 8/4, 10/5. These all reduce to 2/1. It is then said that the scale factor of these two similar triangles is 2 : 1.

The perimeter of $\triangle ABC$ = 24 inches, and the perimeter of $\triangle DEF$ = 12 inches. When you compare the ratios of the perimeters of these similar triangles, you also get 2 : 1. This leads to the following theorem.

Theorem 60: If two similar triangles have a scale factor of *a* : *b*, then the ratio of their perimeters is *a* : *b*.

Example 22: In Figure 155, $\triangle ABC \sim \triangle DEF$. Find the perimeter of $\triangle DEF$.

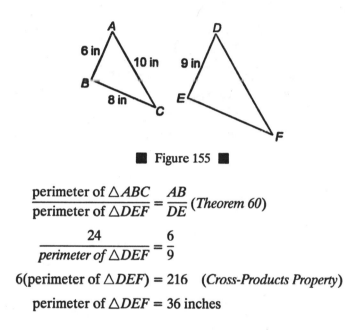

■ Figure 155 ■

$$\frac{\text{perimeter of } \triangle ABC}{\text{perimeter of } \triangle DEF} = \frac{AB}{DE} \ (\textit{Theorem 60})$$

$$\frac{24}{\textit{perimeter of } \triangle DEF} = \frac{6}{9}$$

6(perimeter of $\triangle DEF$) = 216   (*Cross-Products Property*)

perimeter of $\triangle DEF$ = 36 inches

Figure 156 shows two similar right triangles whose scale factor is 2 : 3. Since $\overline{GH} \perp \overline{GI}$ and $\overline{JK} \perp \overline{JL}$, they can be considered base and height for each triangle. You can now find the area of each triangle.

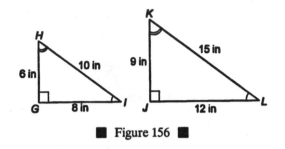

■ Figure 156 ■

area $\triangle GHI = \frac{1}{2}(6)(8)$	area $\triangle JKL = \frac{1}{2}(9)(12)$
area $\triangle GHI = 24$ in²	area $\triangle JKL = 54$ in²

Now you can compare the ratio of the areas of these similar triangles.

$$\frac{\text{area } \triangle GHI}{\text{area } \triangle JKL} = \frac{24}{54}$$

$$= 4/9$$

$$= (2/3)^2$$

This leads to the following theorem.

Theorem 61: If two similar triangles have a scale factor of $a : b$, then the ratio of their areas is $a^2 : b^2$.

Example 23: In Figure 157, $\triangle PQR \sim \triangle STU$. Find the area of $\triangle STU$.

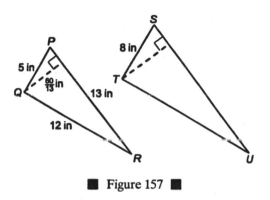

■ Figure 157 ■

The scale of these similar triangles is 5 : 8.

$$\frac{\text{area } \triangle PQR}{\text{area } \triangle STU} = (5/8)^2 \quad (\textit{Theorem 61})$$

$$\text{area } \triangle PQR = \frac{1}{2}(60/13)(13)$$

$$\text{area } \triangle PQR = 30 \text{ in}^2$$

$$\frac{30}{\text{area } \triangle STU} = \frac{25}{64}$$

$$25(\text{area } \triangle STU) = 1920 \quad (\textit{Cross-Products Property})$$

$$\text{area } \triangle STU = 76\frac{4}{5} \text{ in}^2 \text{ or } 76.8 \text{ in}^2$$

Example 24: The perimeters of two similar triangles is in the ratio 3 : 4. The sum of their areas is 75 cm². Find the area of each triangle.

If you call the triangles $\triangle_1$ and $\triangle_2$, then

$$\frac{\text{perimeter } \triangle_1}{\text{perimeter } \triangle_2} = \frac{3}{4}$$

According to *Theorem 60,* this also means that the scale factor of these two similar triangles is 3 : 4.

Let $\qquad\qquad 3x =$ a side in $\triangle_1$

and $\qquad\qquad 4x =$ the corresponding side in $\triangle_2$

Then $\qquad\dfrac{\text{area } \triangle_1}{\text{area } \triangle_2} = \left(\dfrac{3x}{4x}\right)^2$ (*Theorem 61*)

$$\frac{\text{area } \triangle_1}{\text{area } \triangle_2} = \frac{9x^2}{16x^2}$$

Since the sum of the areas is 75 cm², you get

$$\text{area } \triangle_1 + \text{area } \triangle_2 = 9x^2 + 16x^2$$

$$75 = 25x^2$$

$$3 = x^2$$

$$\text{area } \triangle_1 = 9x^2 \qquad \text{area } \triangle_2 = 16x^2$$

$$= 9(3) \qquad\qquad = 16(3)$$

$$= 27 \text{ cm}^2 \qquad\quad = 48 \text{ cm}^2$$

Example 25: The areas of two similar triangles are 45 cm² and 80 cm². The sum of their perimeters is 35 cm. Find the perimeter of each triangle.

Call the two triangles $\triangle_1$ and $\triangle_2$ and let the scale factor of the two similar triangles be $a : b$.

$$\frac{\text{area } \triangle_1}{\text{area } \triangle_2} = \left(\frac{a}{b}\right)^2 \quad \textit{(Theorem 61)}$$

$$\frac{45}{80} = \left(\frac{a}{b}\right)^2$$

Reduce the fraction.
$$\frac{9}{16} = \left(\frac{a}{b}\right)^2$$

Take square roots of both sides.

$$\frac{3}{4} = \frac{a}{b}$$

$a : b$ is the reduced form of the scale factor. 3 : 4 is then the reduced form of the comparison of the perimeters.

Let $\qquad 3x$ = perimeter of $\triangle_1$

and $\qquad 4x$ = perimeter of $\triangle_2$

Then $\quad 3x + 4x = 35 \quad$ (The sum of the perimeters is 35 cm.)

$$7x = 35$$

$$x = 5$$

So $\qquad$ perimeter $\triangle_1 = 3(5) \qquad$ perimeter $\triangle_2 = 4(5)$

$$= 15 \text{ cm} \qquad\qquad\qquad = 20 \text{ cm}$$

Geometric Mean

When a positive value is repeated in either the means or extremes position of a proportion, that value is referred to as a **geometric mean** (or **mean proportional**) between the other two values.

Example 1: Find the geometric mean between 4 and 25.

Let x = the geometric mean.

$$\frac{4}{x} = \frac{x}{25} \quad \text{(definition of geometric mean)}$$

$$x^2 = 100 \quad \text{(Cross-Products Property)}$$

$$x = \sqrt{100}$$

$$x = 10$$

The geometric mean between 4 and 25 is 10.

Example 2: 12 is the geometric mean between 8 and what other value?

Let x = the other value.

$$\frac{8}{12} = \frac{12}{x} \quad \text{(definition of geometric mean)}$$

$$8x = 144 \quad \text{(Cross-Products Property)}$$

$$x = 18$$

The other value is 18.

Altitude to the Hypotenuse

In Figure 158, right triangle *ABC* has altitude $\overline{BD}$ drawn to the hypotenuse $\overline{AC}$.

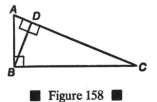

■ Figure 158 ■

The following theorem can now be easily shown using the *AA Similarity Postulate*.

Theorem 62: The altitude drawn to the hypotenuse of a right triangle creates two similar right triangles, each similar to the original right triangle and similar to each other.

Figure 159 shows the three right triangles created in Figure 158. They have been drawn in such a way that corresponding parts are easily recognized.

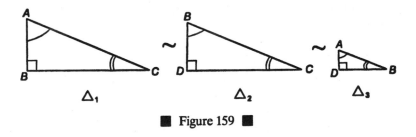

■ Figure 159 ■

Note that $\overline{AB}$ and $\overline{BC}$ are legs of the original right triangle; $\overline{AC}$ is the hypotenuse in the original right triangle; $\overline{BD}$ is the altitude drawn to the hypotenuse; $\overline{AD}$ is the segment on the hypotenuse touching leg $\overline{AB}$; $\overline{DC}$ is the segment on the hypotenuse touching leg $\overline{BC}$.

Because the triangles are similar to one another, ratios of each pair of corresponding sides are equal. This produces three proportions involving geometric means.

(1) $\dfrac{\Delta_1}{\Delta_2}$ $\dfrac{AC}{BC} = \dfrac{BC}{CD}$ $(\Delta_1 \sim \Delta_2)$

(2) $\dfrac{\Delta_1}{\Delta_3}$ $\dfrac{AB}{AD} = \dfrac{AC}{AB}$ $(\Delta_1 \sim \Delta_3)$

These two propotions can now be stated as a theorem.

Theorem 63: If an altitude is drawn to the hypotenuse of a right triangle, then each leg is the geometric mean between the hypotenuse and its touching segment on the hypotenuse.

(3) $\dfrac{\Delta_2}{\Delta_3}$ $\dfrac{BD}{AD} = \dfrac{CD}{BD}$ $(\Delta_2 \sim \Delta_3)$

This proportion can now be stated as a theorem.

Theorem 64: If an altitude is drawn to the hypotenuse of a right triangle, then it is the geometric mean between the segments on the hypotenuse.

Example 3: Use Figure 160 to write three proportions involving geometric means.

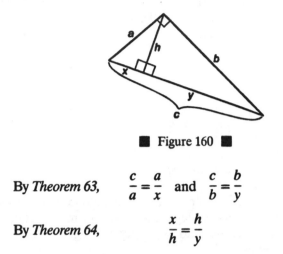

■ Figure 160 ■

By *Theorem 63,* $\dfrac{c}{a} = \dfrac{a}{x}$ and $\dfrac{c}{b} = \dfrac{b}{y}$

By *Theorem 64,* $\dfrac{x}{h} = \dfrac{h}{y}$

Example 4: Find the values for x and y in Figures 161(a) through 161(d).

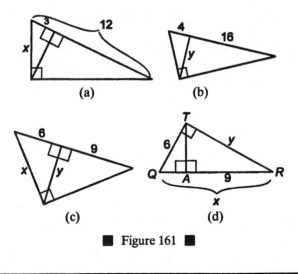

■ Figure 161 ■

(a) By *Theorem 63,*   (b) By *Theorem 64,*

$12/x = x/3$ $\qquad\qquad$ $4/y = y/16$

$\qquad x^2 = 36$ $\qquad\qquad$ $y^2 = 64$

$\qquad x = \sqrt{36}$ $\qquad\qquad$ $y = \sqrt{64}$

$\qquad x = 6$ $\qquad\qquad$ $y = 8$

(c) By *Theorem 63,* $\qquad$ By *Theorem 64,*

$15/x = x/6$ $\qquad\qquad$ $6/y = y/9$

$\qquad x^2 = 90$ $\qquad\qquad$ $y^2 = 54$

$\qquad x = \sqrt{90}$ $\qquad\qquad$ $y = \sqrt{54}$

$\qquad x = (\sqrt{9})(\sqrt{10})$ $\qquad$ $y = (\sqrt{9})(\sqrt{6})$

$\qquad x = 3\sqrt{10}$ $\qquad\qquad$ $y = 3\sqrt{6}$

(d) $QA + AR = QR$ $\quad$ (*Segment Addition Postulate*)

$\qquad QA + 9 = x$

$\qquad QA = x - 9$

By *Theorem 63,* $\;x/6 = 6/(x - 9)$

$\qquad x(x - 9) = 36$ $\quad$ (*Cross-Products Property*)

$\qquad x^2 - 9x = 36$

$\qquad x^2 - 9x - 36 = 0$

Factor, $\qquad (x - 12)(x + 3) = 0$

$\qquad x - 12 = 0$ $\quad$ or $\quad x + 3 = 0$

$\qquad x = 12$ $\quad$ or $\qquad x = -3$

Since it represents a length, x cannot be negative, so $x = 12$.

By *Theorem 64,* $x/y = y/9$

Since $x = 12$, from above,

$$12/y = y/9 \quad (\text{*Cross-Products Property*})$$

$$y^2 = 108$$

$$y = \sqrt{108}$$

$$y = (\sqrt{36})(\sqrt{3})$$

$$y = 6\sqrt{3}$$

Pythagorean Theorem and Its Converse

In Figure 162, $\overline{CD}$ is the altitude to hypotenuse $\overline{AB}$.

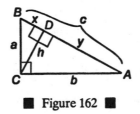

■ Figure 162 ■

So, by *Theorem 63,*

$$c/a = a/x, \quad \text{which becomes} \quad a^2 = cx$$

and $\qquad c/b = b/y, \quad \text{which becomes} \quad b^2 = cy$

From the addition property of equations in algebra, we get the following equation.

$$a^2 + b^2 = cx + cy$$

By factoring out the c on the right side,

$$a^2 + b^2 = c\underbrace{(x + y)}$$

But $x + y = c$ (*Segment Addition Postulate*),

$$a^2 + b^2 = cc \quad \text{or} \quad a^2 + b^2 = c^2$$

This result is known as the *Pythagorean Theorem*.

Theorem 65: (*Pythagorean Theorem*) **In any right triangle, the sum of the squares of the legs equals the square of the hypotenuse (leg² + leg² = hypotenuse²).**

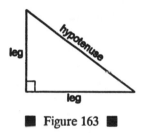

■ Figure 163 ■

Example 5: In Figure 164, find x, the length of the hypotenuse.

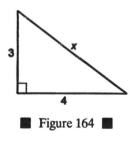

■ Figure 164 ■

$$\text{leg}^2 + \text{leg}^2 = \text{hypotenuse}^2$$
$$3^2 + 4^2 = x^2$$
$$9 + 16 = x^2$$

$$25 = x^2$$

$$\sqrt{25} = x$$

$$5 = x$$

Example 6: Use Figure 165 to find x.

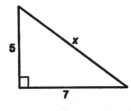

■ Figure 165 ■

$$\text{leg}^2 + \text{leg}^2 = \text{hypotenuse}^2$$

$$5^2 + 7^2 = x^2$$

$$25 + 49 = x^2$$

$$74 = x^2$$

$$\sqrt{74} = x$$

Any three natural numbers, a, b, c, that make the sentence $a^2 + b^2 = c^2$ true are called **Pythagorean triples.** Therefore, 3-4-5 is called a Pythagorean triple. Some other values for a, b, and c that will work are 5-12-13 and 8-5-17. Any multiple of one of these triples will also work. For example, using the 3-4-5: 6-8-10, 9-12-15, and 15-20-25 are also Pythagorean triples.

Example 7: Use Figure 166 to find x.

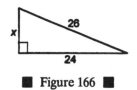

■ Figure 166 ■

If you can recognize that the numbers x, 24, 26 are a multiple of the 5-12-13 Pythagorean triple, the answer for x is quickly found. Since $24 = 2(12)$ and $26 = 2(13)$, then $x = 2(5)$ or $x = 10$. You can also find x by using the *Pythagorean Theorem*.

$$\text{leg}^2 + \text{leg}^2 = \text{hypotenuse}^2$$
$$x^2 + 24^2 = 26^2$$
$$x^2 + 576 = 676$$
$$x^2 = 100$$
$$x = \sqrt{100}$$
$$x = 10$$

Example 8: Use Figure 167 to find x.

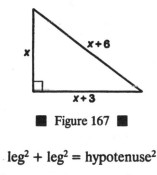

■ Figure 167 ■

$$\text{leg}^2 + \text{leg}^2 = \text{hypotenuse}^2$$
$$x^2 + (x + 3)^2 = (x + 6)^2$$

$$x^2 + x^2 + 6x + 9 = x^2 + 12x + 36$$

$$2x^2 + 6x + 9 = x^2 + 12x + 36$$

Subtract $x^2 + 12x + 36$ from both sides.

$$x^2 - 6x - 27 = 0$$

Factor. $(x - 9)(x + 3) = 0$

$$x - 9 = 0 \quad \text{or} \quad x + 3 = 0$$

So $x = 9 \quad \text{or} \quad x = -3$

But x is a length, so it cannot be negative. Therefore, $x = 9$.

The converse (reverse) of the *Pythagorean Theorem* is also true.

Theorem 66: If a triangle has sides of lengths a, b, and c where c is the longest length and $c^2 = a^2 + b^2$, then the triangle is a right triangle with c its hypotenuse.

Example 9: Determine if the following sets of lengths could be the sides of a right triangle: (a) 6-5-4, (b) $\sqrt{11}$-$\sqrt{14}$-5, (c) 3/4-1-5/4

(a) Since 6 is the longest length, do the following check.

$$6^2 ? 4^2 + 5^2$$

$$36 ? 16 + 25$$

$$36 \neq 41$$

So 4-5-6 are not the sides of a right triangle.

(b) Since 5 is the longest length, do the following check.

$$5^2 ? (\sqrt{11})^2 + (\sqrt{14})^2$$

$$25 \; ? \; 11 + 14$$

$$25 = 25$$

So $\sqrt{11}$-$\sqrt{14}$-5 are sides of a right triangle, and 5 is the length of the hypotenuse.

(c) Since 5/4 is the longest length, do the following check.

$$(5/4)^2 \; ? \; (3/4)^2 + (1)^2$$

$$25/16 \; ? \; 9/16 + 1$$

$$25/16 = 25/16$$

So 3/4-1-5/4 are sides of a right triangle, and 5/4 is the length of the hypotenuse.

Extension to the Pythagorean Theorem

Variations of *Theorem 66* can be used to classify a triangle as right, obtuse, or acute.

Theorem 67: If a, b, and c represent the lengths of the sides of a triangle, and c is the longest length, then the triangle is obtuse if $c^2 > a^2 + b^2$, and the triangle is acute if $c^2 < a^2 + b^2$.

Figures 168(a) through 168(c) show these different triangle situations and the sentences comparing their sides. In each case, c represents the longest side in the triangle.

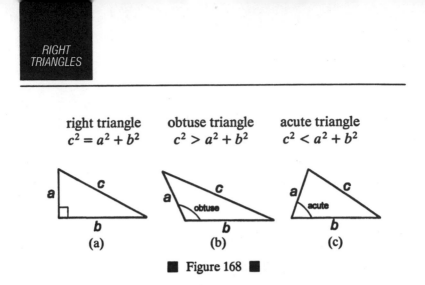

right triangle   obtuse triangle   acute triangle
$c^2 = a^2 + b^2$ $c^2 > a^2 + b^2$ $c^2 < a^2 + b^2$

(a)   (b)   (c)

■ Figure 168 ■

Example 10: Determine whether the following sets of three values could be the lengths of the sides of a triangle. If the values can be the sides of a triangle, then classify the triangle.
(a) 16-30-34, (b) 5-5-8, (c) 5-8-15, (d) 4-4-5, (e) 9-12-16, (f) 1-1-$\sqrt{2}$

(Recall the *Triangle Inequality Theorem, Theorem 38,* which states that the longest side in any triangle must be less than the sum of the two shorter sides.)

(a)   34 ? 16 + 30

        34 < 46   (So these can be the sides of a triangle.)

        34^2 ? $16^2 + 30^2$

    1156 ? 256 + 900

    1156 = 1156

    This is a right triangle. Since its sides are of different lengths, it is also a scalene triangle.

(b)   8 ? 5 + 5

       8 < 10   (So these can be the sides of a triangle.)

       8^2 ? $5^2 + 5^2$

       64 ? 25 + 25

       64 > 50

This is an obtuse triangle. Since two of its sides are of equal measure, it is also an isosceles triangle.

(c) 15 ? 5 + 8

15 > 13 (So these *cannot* be the sides of a triangle.)

(d) 5 ? 4 + 4

5 < 8 (So these can be the sides of a triangle.)

5^2 ? 4^2 + 4^2

25 ? 16 + 16

25 < 32

This is an acute triangle. Since two of its sides are of equal measure, it is also an isosceles triangle.

(e) 16 ? 9 + 12

16 < 21 (So these can be the sides of a triangle.)

16^2 ? 9^2 + 12^2

256 ? 81 + 144

256 > 225

This is an obtuse triangle. Since all sides are of different lengths, it is also a scalene triangle.

(f) $\sqrt{2}$? 1 + 1

$\sqrt{2}$ < 2 (So these can be the sides of a triangle.)

$\sqrt{2}^2$? 1^2 + 1^2

2 ? 1 + 1

2 = 2

This is a right triangle. Since two of its sides are of equal measure, it is also an isosceles triangle.

Special Right Triangles

Isosceles right triangle. An **isosceles right triangle** has the characteristic of both the isosceles and the right triangles. It has two equal sides, two equal angles, and one right angle. (The right angle cannot be one of the equal angles or the sum of the angles would exceed 180°). Therefore, in Figure 169, $\triangle ABC$ is an isosceles right triangle, and the following must always be true.

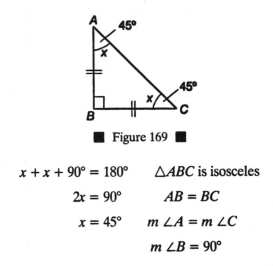

■ Figure 169 ■

$$x + x + 90° = 180° \qquad \triangle ABC \text{ is isosceles}$$
$$2x = 90° \qquad AB = BC$$
$$x = 45° \qquad m \angle A = m \angle C$$
$$m \angle B = 90°$$

The ratio of the sides of an isosceles right triangle is always $1 : 1 : \sqrt{2}$ or $x : x : x\sqrt{2}$.

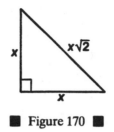

■ Figure 170 ■

Example 11: If one of the equal sides of an isosceles triangle is 3, what is the measure of the other two sides?

Method 1: Using the ratio $x : x : x\sqrt{2}$ for isosceles right triangles, then $x = 3$, and the other sides must be 3 and $3\sqrt{2}$.

Method 2: Using the *Pythagorean Theorem* and the fact that the legs of this right triangle are equal,

$$\text{leg}^2 + \text{leg}^2 = \text{hypotenuse}^2$$

$$3^2 + 3^2 = \text{hypotenuse}^2$$

$$9 + 9 = \text{hypotenuse}^2$$

$$18 = \text{hypotenuse}^2$$

So $$\text{hypotenuse} = \sqrt{18}$$

or $$\text{hypotenuse} = 3\sqrt{2}$$

The two sides have measures of 3 and $3\sqrt{2}$.

Example 12: If the diagonal of a square is $6\sqrt{2}$, find the length of each of its sides.

Method 1: The diagonal of a square divides it into two congruent isosceles right triangles. Look at Figure 171.

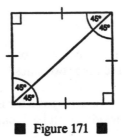

■ Figure 171 ■

The ratio $x : x : x\sqrt{2}$ for isosceles right triangles can now be applied where $x\sqrt{2} = 6\sqrt{2}$. So $x = 6$, and each side of the square has a measure of 6.

Method 2: Use the *Pythagorean Theorem.* $6\sqrt{2}$ represents the hypotenuse.

$$\text{leg}^2 + \text{leg}^2 = \text{hypotenuse}^2$$
$$(2)\text{leg}^2 = (6\sqrt{2})^2$$
$$(2)\text{leg}^2 = (36)(2)$$
$$(2)\text{leg}^2 = 72$$
$$\text{leg}^2 = 36$$
$$\text{leg} = 6$$

Therefore, each side of the square has a measure of 6.

Example 13: What are the measurements of x, y, and z in Figure 172?

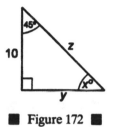

■ Figure 172 ■

$45° + 90° + x° = 180°$ (The sum of the angles of a triangle = 180°.)

$x = 45°$

Therefore, this is an isosceles right triangle with the ratio of sides x : x : $x\sqrt{2}$. Since one leg is 10, the other must also be 10, and the hypotenuse is $10\sqrt{2}$, so $y = 10$ and $z = 10\sqrt{2}$.

30°-60°-90° right triangle. A **30°-60°-90° right triangle** has a unique ratio of its sides. The ratio of the sides of a 30°-60°-90° right triangle is $1 : \sqrt{3} : 2$ or $x : x\sqrt{3} : 2x$, placed as follows.

The side opposite 30° is the shortest side and is 1 or x (Figure 173).
The side opposite 60° is $\sqrt{3}$ or $x\sqrt{3}$.
The side opposite 90° is the longest side (hypotenuse) and is 2 or $2x$.

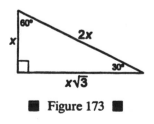

■ Figure 173 ■

Example 14: If the shortest side of a 30°-60°-90° right triangle is 4, what is the measure of the other two sides?

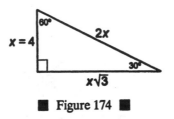

■ Figure 174 ■

In Figure 174, x is opposite the 30°. The other two sides are then $x\sqrt{3}$ (opposite the 60°) and $2x$ (opposite the 90°). Since the shortest side is 4, $x = 4$. Then the other two sides must be $4\sqrt{3}$ and 2(4), or 8.

Example 15: If the longer leg of a 30°-60°-90° right triangle is $8\sqrt{3}$, find the length of the hypotenuse.

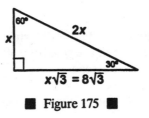

■ Figure 175 ■

In Figure 175, the shorter leg, x, is opposite the 30°. $x\sqrt{3}$ is the longer leg, and it is opposite the 60°. The hypotenuse is $2x$. Since $x\sqrt{3} = 8\sqrt{3}$, $x = 8$. Since $x = 8$, then $2x = 16$. The hypotenuse is 16.

Example 16: Find the length of an altitude in an equilateral triangle with a perimeter of 60 inches.

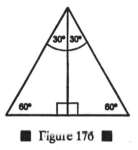

■ Figure 176 ■

Figure 176 is an equilateral triangle. Each angle has a measure of 60°. If an altitude is drawn, it creates two 30°-60°-90° right triangles. Since the perimeter is 60 inches, and the three sides are equal in measure, then each side is 20 inches (60 ÷ 3 = 20). The ratio of sides in a 30°-60°-90° right triangle is $x : x\sqrt{3} : 2x$. In this problem, the length 20 inches represents the longest side in the 30°-60°-90° right triangle, so $2x = 20$, or $x = 10$. Since the altitude is the longer leg of the 30°-60°-90° right triangle and its measure is $x\sqrt{3}$, the altitude is $10\sqrt{3}$ inches long.

Parts of Circles

- **Circle.** To review, a **circle** is a plane figure with all points the same distance from one point.

- **Center.** That one point is called the **center** of the circle. Circles are named by naming the center.

- **Radius.** Any segment whose endpoints lie one at the center of the circle and one on the circle is a **radius.** (The plural of *radius* is *radii.*)

- **Chord.** Any segment whose endpoints lie on the circle is a **chord.**

- **Diameter.** Any chord that passes through the center of the circle is a **diameter.**

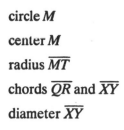

circle *M*

center *M*

radius $\overline{MT}$

chords $\overline{QR}$ and $\overline{XY}$

diameter $\overline{XY}$

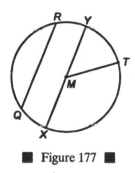

■ Figure 177 ■

From the definition of radius and diameter, it is clear that all radii of a circle are equal in length and all diameters of a circle are equal in length.

■ **Secant.** Any line that contains a chord is a **secant.**

■ **Tangent.** Any line in the same plane as a circle and that intersects the circle in exactly one point is a **tangent.**

■ **Point of tangency.** The point where a tangent line intersects a circle is the **point of tangency.**

circle Q

$\overleftrightarrow{AB}$ is a secant.

$\overleftrightarrow{CE}$ is a tangent.

D is the point of tangency for $\overleftrightarrow{CE}$.

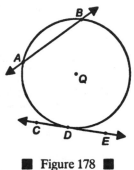

■ Figure 178 ■

■ **Common tangents.** A line that is tangent to two circles in the same plane is a **common tangent.**

■ **Internal common tangent.** A common tangent that intersects the segment joining the centers of two circles is an **internal common tangent.**

■ **External common tangent.** A common tangent that does not intersect the segment joining the centers of two circles is an **external common tangent.** In Figure 179,

Lines *l* and *m* are common tangents.

l is an internal common tangent.

m is an external common tangent.

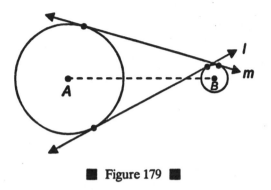

■ Figure 179 ■

Example 1: Use Figure 180 to find each of the following.

(a) a radius in circle O
(b) a chord in circle P
(c) a diameter in circle O
(d) a secant in circle P
(e) a tangent to circle O
(f) a common internal tangent to circles O and P
(g) a common external tangent to circles O and P

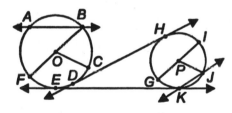

■ Figure 180 ■

(a) $\overline{OB}, \overline{OC},$ or $\overline{OF}$ (b) $\overline{GI}$ or $\overline{JK}$ (c) $\overline{BF}$ (d) $\overleftrightarrow{JK}$

(e) $\overleftrightarrow{KE}$ or $\overleftrightarrow{DH}$ (f) $\overleftrightarrow{DH}$ (g) $\overleftrightarrow{EK}$

Arcs and Central Angles

- **Central angles. Central angles** are angles formed by any two radii in a circle. The vertex is the center of the circle. In Figure 181, $\angle AOB$ is a central angle.

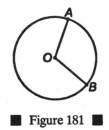

■ Figure 181 ■

- **Arc.** An **arc** of a circle is a portion of the circle. It includes two endpoints and all the points on the circle between these endpoints. The symbol ⌒ is used to denote an arc. This symbol is written over the endpoints that form the arc. There are three types of arcs.

- **Semicircle.** A **semicircle** is an arc whose endpoints are the endpoints of a diameter. It is named using three points. The first and third points are the endpoints of the diameter, and the middle point is any point of the arc between the endpoints.

- **Minor arc.** A **minor arc** is an arc which is less than a semicircle. A minor arc is named by using only the two endpoints of the arc.

- **Major arc.** A **major arc** is an arc which is more than a semicircle. It is named by three points. The first and third are the endpoints, and the middle point is any point on the arc between the endpoints.

In Figure 182, $\overline{AC}$ is a diameter.

$\overparen{ABC}$ is a semicircle.

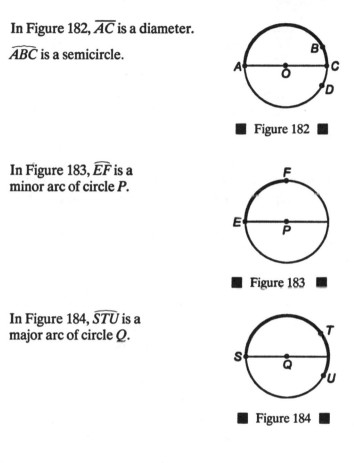

■ Figure 182 ■

In Figure 183, $\overparen{EF}$ is a minor arc of circle P.

■ Figure 183 ■

In Figure 184, $\overparen{STU}$ is a major arc of circle Q.

■ Figure 184 ■

Arcs are measured in two different ways. They are measured in degrees and in unit length.

■ **Degree measure of a semicircle.** The **degree measure of a semicircle** is 180°. Its unit length is half of the circumference of the circle.

■ **Degree measure of a minor arc.** The **degree measure of a minor arc** is defined to be the same as the measure of its corresponding central angle. Its unit length is a portion of the circumference. Its length is always less than half of the circumference.

- **Degree measure of a major arc.** The **degree measure of a major arc** is 360° minus the degree measure of its corresponding central angle. Its unit length is a portion of the circumference and is always more than half of the circumference.

In this book, $m\ \overset{\frown}{AB}$ indicates the degree measure of arc AB, $l\ \overset{\frown}{AB}$ indicates the length of arc AB, and $\overset{\frown}{AB}$ indicates the arc itself.

Example 2: In Figure 185, circle O, with diameter $\overline{AB}$, has $OB = 6$ inches. Find (a) $m\ \overset{\frown}{AXB}$ and (b) $l\ \overset{\frown}{AXB}$.

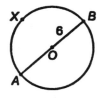

■ Figure 185 ■

(a) $\overset{\frown}{AXB}$ is a semicircle. $m\ \overset{\frown}{AXB} = 180°$.

(b) Since $\overset{\frown}{AXB}$ is a semicircle, its length is half of the circumference.

$$l\ \overset{\frown}{AXB} = \tfrac{1}{2}(2\pi r)$$

$$l\ \overset{\frown}{AXB} = \tfrac{1}{2}(2\pi 6 \text{ inches})$$

$$l\ \overset{\frown}{AXB} = 6\pi \text{ inches}$$

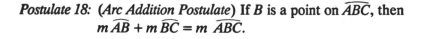

Postulate 18: (*Arc Addition Postulate*) If *B* is a point on $\overset{\frown}{ABC}$, then
$m\,\overset{\frown}{AB} + m\,\overset{\frown}{BC} = m\,\overset{\frown}{ABC}$.

Example 3: Use Figure 186 to find $m\,\overset{\frown}{ABC}$ ($m\,\overset{\frown}{AB} = 60°$,
$m\,\overset{\frown}{BC} = 150°$).

By *Postulate 18,*

$m\,\overset{\frown}{ABC} = m\,\overset{\frown}{AB} + m\,\overset{\frown}{BC}$

$m\,\overset{\frown}{ABC} = 60° + 150°$

$m\,\overset{\frown}{ABC} = 210°$

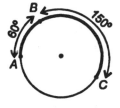

■ Figure 186 ■

Example 4: Use Figure 187 of circle *P* with diameter $\overline{QS}$ to answer
the following

(a) Find $m\,\overset{\frown}{RS}$.

(b) Find $m\,\overset{\frown}{QRS}$.

(c) Find $m\,\overset{\frown}{QR}$.

(d) Find $m\,\overset{\frown}{RQS}$.

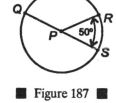

■ Figure 187 ■

(a) $m\,\overset{\frown}{RS} = 50°$ (The degree measure of a minor arc equals the
measure of its corresponding central angle.)

(b) $m\,\overset{\frown}{QRS} = 180°$ ($\overset{\frown}{QRS}$ is a semicircle.)

(c) $m\,\widehat{QR} = 130°$

$$m\,\widehat{QR} + m\,\widehat{RS} = m\,\widehat{QRS} \text{ (by } Postulate\ 18)$$

$$m\,\widehat{QR} = m\,\widehat{QRS} - m\,\widehat{RS}$$

$$m\,\widehat{QR} = 180° - 50° \text{ (or } 130°)$$

(d) $m\,\widehat{RQS} = 310°$ ($\widehat{RQS}$ is a major arc. The degree measure of a major arc is 360° minus its corresponding central angle.)

$$m\,\widehat{RQS} = 360° - m\,\widehat{RS}$$

$$m\,\widehat{RQS} = 360° - 50° \text{ (or } 310°)$$

The following theorems about arcs and central angles are easily proven.

Theorem 68: In a circle, if two central angles have equal measures, then their corresponding minor arcs have equal measures.

Theorem 69: In a circle, if two minor arcs have equal measures, then their corresponding central angles have equal measures.

Example 5: Figure 188 shows circle O with diameters $\overline{AC}$ and $\overline{BD}$. If $m\,\angle 1 = 40°$, find each of the following.

(a) $m\,\widehat{AB}$

(b) $m\,\widehat{CD}$

(c) $m\,\widehat{AD}$

(d) $m\,\angle DOA$

(e) $m\,\angle 3$

(f) $m\,\angle 4$

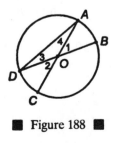

■ Figure 188 ■

(a) $m\,\widehat{AB} = 40°$ (The measure of a minor arc equals the measure of its corresponding central angle.)

(b) $m\,\widehat{CD} = 40°$ (Since vertical angles have equal measures, $m\,\angle 1 = m\,\angle 2$. Then the measure of a minor arc equals the measure of its corresponding central angle.)

(c) $m\,\widehat{AD} = 140°$ (By *Postulate 18*, $m\,\widehat{AD} + m\,\widehat{AB} = m\,\widehat{DAB}$; $\widehat{DAB}$ is a semicircle, so $m\,\widehat{AD} + 40° = 180°$, or $m\,\widehat{AD} = 140°$.)

(d) $m\,\angle DOA = 140°$ (The measure of a central angle equals the measure of its corresponding minor arc.)

(e) $m\,\angle 3 = 20°$ (Since radii of a circle are equal, $OD = OA$. Since, if two sides of a triangle are equal, then the angles opposite these sides are equal, $m\,\angle 3 = m\,\angle 4$. Since the sum of the angles of any triangle equals 180°, $m\,\angle 3 + m\,\angle 4 + m\,\angle DOA = 180°$. By replacing $m\,\angle 4$ with $m\,\angle 3$ and $m\,\angle DOA$ with 140°,

$$m\,\angle 3 + m\,\angle 3 + 140° = 180°$$

or $\qquad\qquad\qquad 2(m\,\angle 3) = 40°$

or $\qquad\qquad\qquad m\,\angle 3 = 20°$

(f) $m\,\angle 4 = 20°$ (As discussed above, $m\,\angle 3 = m\,\angle 4$.)

Arcs and Inscribed Angles

- **Inscribed angle.** An **inscribed angle** in a circle is an angle formed by two chords with the vertex on the circle.

- **Intercepted arc.** The **intercepted arc** corresponding to an angle is the portion of the circle that lies in the interior of the angle including the endpoints of the arc.

In Figure 189, ∠*ABC* is an inscribed angle and $\overset{\frown}{AC}$ is its intercepted arc.

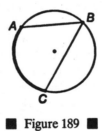

■ Figure 189 ■

Figure 190 shows examples of angles that are *not* inscribed angles.

∠*QRS* is *not* an inscribed angle, since its vertex is not on the circle.   ∠*TWV* is *not* an inscribed angle, since its vertex is not on the circle.

■ Figure 190 ■

Refer to Figure 188 and the example that accompanies it. Notice that $m \angle 4$ is exactly half of $m \overarc{AB}$, and $m \angle 3$ is half of $m \overarc{CD}$. $\angle 3$ and $\angle 4$ are inscribed angles, and $\overarc{AB}$ and $\overarc{CD}$ are their intercepted arcs, which leads to the following theorem.

Theorem 70: The measure of an inscribed angle in a circle equals half the measure of its intercepted arc.

The following two theorems directly follow from *Theorem 70.*

Theorem 71: If two inscribed angles of a circle intercept the same arc or arcs of equal measure, then the inscribed angles have equal measures.

Theorem 72: If an inscribed angle intercepts a semicircle, then its measure is 90°.

Example 6: Find $m \angle C$ in Figure 191.

$m \angle C = \frac{1}{2}(m \overarc{BD})$ (*Theorem 70*)

$m \angle C = \frac{1}{2}(60°)$

$m \angle C = 30°$

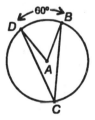

■ Figure 191 ■

Example 7: Find $m \angle A$ and $m \angle B$ in Figure 192.

$m \angle A = \frac{1}{2}(m \overarc{CD})$ (*Theorem 70*)

$m \angle A = \frac{1}{2}(110°)$

$m \angle A = 55°$

$m \angle B = 55°$ (*Theorem 71*)

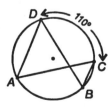

■ Figure 192 ■

Example 8: In Figure 193, $\overline{QS}$ is a diameter. Find $m \angle R$.

$m \angle R = 90°$ *(Theorem 72)*

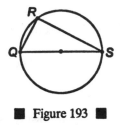

■ Figure 193 ■

Example 9: In Figure 194 of circle O, $m \widehat{CD} = 60°$ and $m \angle 1 = 25°$. Find each of the following.

(a) $m \angle CAD$

(b) $m \widehat{BC}$

(c) $m \angle BOC$

(d) $m \widehat{AB}$

(e) $m \angle ACB$

(f) $m \angle ABC$

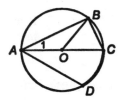

■ Figure 194 ■

(a) $m \angle CAD = \frac{1}{2}(m \widehat{CD})$ *(Theorem 70)*

$m \angle CAD = \frac{1}{2}(60°)$

$m \angle CAD = 30°$

(b) $m \angle 1 = \frac{1}{2}(m \widehat{BC})$ *(Theorem 70)*

Multiply each side by 2.

$$2(m \angle 1) = m \widehat{BC}$$
$$2(25°) = m \widehat{BC}$$
$$m \widehat{BC} = 50°$$

(c) $m \angle BOC = 50°$ (The measure of a central angle equals the measure of its corresponding minor arc.)

(d) $m \overset{\frown}{AB} + m \overset{\frown}{BC} = m \overset{\frown}{ABC}$ (*Arc Addition Postulate*)

$$m \overset{\frown}{AB} + 50° = 180°$$
$$m \overset{\frown}{AB} = 130°$$

(e) $m \angle ACB = \frac{1}{2}(m \overset{\frown}{AB})$ (*Theorem 70*)

$$m \angle ACB = \frac{1}{2}(130°)$$
$$m \angle ACD = 65°$$

(f) $m \angle ABC = 90°$ (*Theorem 72*)

Other Angles Formed by Chords, Secants, and/or Tangents

Theorem 73: If a tangent and a diameter meet at the point of tangency, then they are perpendicular to one another.

In Figure 195, diameter $\overline{AB}$ meets tangent $\overleftrightarrow{CD}$ at B. According to Theorem 73, $\overline{AB} \perp \overleftrightarrow{CD}$, which means that $m \angle ABC = 90°$ and $m \angle ABD = 90°$.

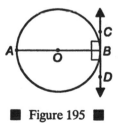

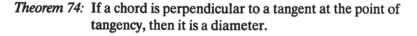
■ Figure 195 ■

Theorem 74: If a chord is perpendicular to a tangent at the point of tangency, then it is a diameter.

Example 10: *Theorem 74* could be used to find the center of a circle if two tangents to the circle were known. In Figure 196, $\overleftrightarrow{MN}$ is tangent to the circle at *P*, and $\overleftrightarrow{QR}$ is tangent to the circle at *S*. Use these facts to find the center of the circle.

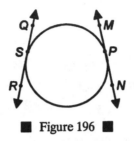

■ Figure 196 ■

According to *Theorem 74,* if a chord is drawn perpendicular to $\overleftrightarrow{MN}$ at *P*, it is a diameter, which means that it passes through the center of the circle. Similarly, if a chord is drawn perpendicular to $\overleftrightarrow{QR}$ at *S*, it too would be a diameter and pass through the center of the circle. The point where these two chords intersect would then be the center of the circle. See Figure 197.

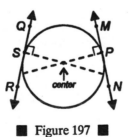

■ Figure 197 ■

Theorem 75: The measure of an angle formed by two chords intersecting inside a circle is equal to half the sum of the measures of the intecepted arcs.

In Figure 198, chords $\overline{AC}$ and $\overline{BD}$ intersect inside the circle at *E*. By *Theorem 75,*

$$m \angle 1 = \tfrac{1}{2}(m\,\widehat{AB} + m\,\widehat{CD})$$

and

$$m \angle 2 = \tfrac{1}{2}(m\,\widehat{AD} + m\,\widehat{BC})$$

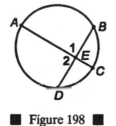

■ Figure 198 ■

Theorem 76: The measure of an angle formed by a tangent and a chord meeting at the point of tangency is half the measure of the intercepted arc.

In Figure 199, chord $\overline{QR}$ and tangent $\overleftrightarrow{TS}$ meet at R. By *Theorem 76*, $m \angle 1 = \tfrac{1}{2}(m\,\widehat{QR})$ and $m \angle 2 = \tfrac{1}{2}(m\,\widehat{QMR})$.

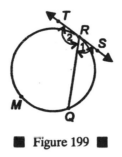

■ Figure 199 ■

Theorem 77: The measure of an angle formed by two secants intersecting outside a circle is equal to one-half the difference of the measures of the intercepted arcs.

In Figure 200, secants $\overleftrightarrow{EF}$ and $\overleftrightarrow{IH}$ intersect at *G*. According to *Theorem 77*, $m \angle 1 = \frac{1}{2}(m \widehat{EI} - m \widehat{FH})$.

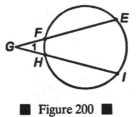

■ Figure 200 ■

Example 11: Find $m \angle 1$ in Figures 201(a) through 201(d).

(a) $m \angle 1 = 90°$ (*Theorem 73*)

(b) $m \angle 1 = \frac{1}{2}(45° + 70°)$ (*Theorem 75*)
      $m \angle 1 = \frac{1}{2}(115°)$
      $m \angle 1 = 57\frac{1}{2}°$, or 57.5°

(c) $m \angle 1 = \frac{1}{2}(228°)$ (*Theorem 76*)
      $m \angle 1 = 114°$

(d) $m \angle 1 = \frac{1}{2}(98° - 33°)$ (*Theorem 77*)
      $m \angle 1 = \frac{1}{2}(65°)$
      $m \angle 1 = 32\frac{1}{2}°$, or 32.5°

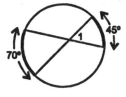
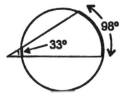
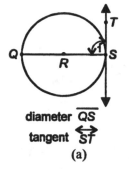
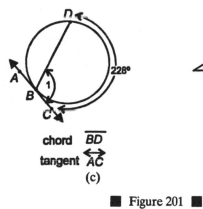

diameter $\overline{QS}$
tangent $\overleftrightarrow{ST}$

(a)

(b)

chord $\overline{BD}$
tangent $\overleftrightarrow{AC}$

(c)

(d)

■ Figure 201 ■

Example 12: Find the value of *y* in each of the following figures.

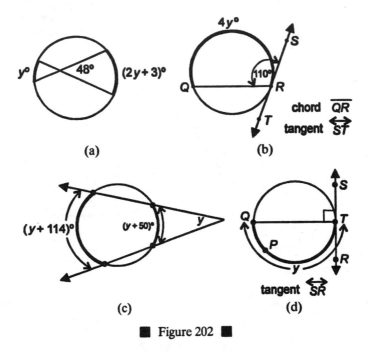

(a)

(b)

(c)

(d)

■ Figure 202 ■

(a) $48 = \frac{1}{2}(y + 2y + 3)$ (*Theorem 75*)

Multiply each side by 2 and simplify.

$96 = 3y + 3$

$93 = 3y$

$31° = y$

(b) $110 = \frac{1}{2}(4y)$ (*Theorem 76*)

$110 = 2y$

$55° = y$

(c) $y = \frac{1}{2}[(y + 114) - (y + 50)]$ (*Theorem 77*)

Multiply each side by 2 and simplify.

$2y = y + 114 - y - 50$

$2y = 64$

$y = 32°$

(d) $y = 180°$ (According to *Theorem 74*, $\overline{QT}$ is a diameter, which would make $\overarc{QPT}$ a semicircle.)

Arcs and Chords

In Figure 203, circle O has radii $\overline{OA}, \overline{OB}, \overline{OC}$ and $\overline{OD}$. If chords $\overline{AB}$ and $\overline{CD}$ are of equal length, it can be shown that $\triangle AOB \cong \triangle DOC$. This would make $m \angle 1 = m \angle 2$, which in turn would make $m \overarc{AB} = m \overarc{CD}$. This is stated as a theorem.

Theorem 78: In a circle, if two chords are equal in measure, then their corresponding minor arcs are equal in measure.

The converse of this theorem is also true.

Theorem 79: In a circle, if two minor arcs are equal in measure, then their corresponding chords are equal in measure.

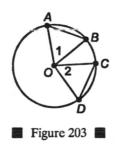

■ Figure 203 ■

Example 13: Use Figure 204 to determine the following. (a) If $AB = CD$, and $m\ \widehat{AB} = 60°$, find $m\ CD$. (b) If $m\ \widehat{EF} = m\ \widehat{GH}$, and $EF = 8$, find GH.

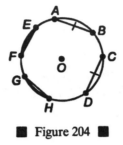

■ Figure 204 ■

(a) $m\ \widehat{CD} = 60°$ (*Theorem 78*)

(b) $GH = 8$ (*Theorem 79*)

Some additional theorems about chords in a circle are presented below without explanation. These theorems can be used to solve many types of problems.

Theorem 80: If a diameter is perpendicular to a chord, then it bisects the chord and its arcs.

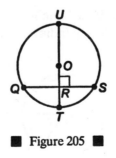

■ Figure 205 ■

In Figure 205, diameter $\overline{UT}$ is perpendicular to chord $\overline{QS}$. By *Theorem 80*, $QR = RS$, $m\,\overset{\frown}{QT} = m\,\overset{\frown}{ST}$, and $m\,\overset{\frown}{QU} = m\,\overset{\frown}{SU}$.

Theorem 81: In a circle, if two chords are equal in measure, then they are equidistant from the center.

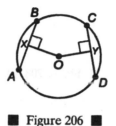

■ Figure 206 ■

In Figure 206, if $AB = CD$, then by *Theorem 81*, $OX = OY$.

Theorem 82: In a circle, if two chords are equidistant from the center of a circle, then the two chords are equal in measure.

In Figure 206, if $OX = OY$, then by *Theorem 82*, $AB = CD$.

Example 14: Use Figure 207 to find x.

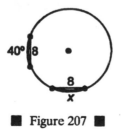

■ Figure 207 ■

$x = 40°$ (*Theorem 78*)

Example 15: Use Figure 208, in which $m\,\widehat{AC} = 115°$, $m\,\widehat{BD} = 115°$, and $BD = 10$, to find AC.

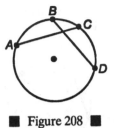

■ Figure 208 ■

$AC = 10$ (*Theorem 79*)

Example 16: Use Figure 209, in which $AB = 10$, $OA = 13$, and $m\,\angle AOB = 55°$, to find OM, $m\,\widehat{AT}$, and $m\,\widehat{SB}$.

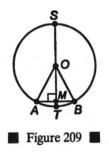

■ Figure 209 ■

Since $\overline{ST} \perp \overline{AB}$, and $\overline{ST}$ is a diameter. *Theorem 80* says that $AM = BM$. Since $AB = 10$, then $AM = 5$. Now consider right triangle AMO. Since $OA = 13$ and $AM = 5$, OM can be found by using the *Pythagorean Theorem.*

$$OM^2 + AM^2 = AQ^2$$

$$OM^2 + 25 = 169$$

$$OM^2 = 144$$

$$OM = \sqrt{144}$$

$$OM = 12$$

Also, *Theorem 80* says that $m\,\widehat{AT} = m\,\widehat{TB}$ and $m\,\widehat{AS} = m\,\widehat{SB}$. Since $m \angle AOB = 55°$, that would make $m\,\widehat{AB} = 55°$ and $m\,\widehat{ASB} = 305°$. Therefore, $m\,\widehat{AT} = 27\frac{1}{2}°$ and $m\,\widehat{SB} = 152\frac{1}{2}°$.

Example 17: Use Figure 210, in which $AB = 8$, $CD = 8$, and $OA = 5$, to find ON.

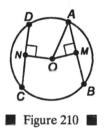

■ Figure 210 ■

By *Theorem 81*, $ON = OM$. By *Theorem 80*, $AM = MB$, so $AM = 4$. OM can now be found by the use of the *Pythagorean Theorem* or by recognizing a Pythagorean triple. In either case, $OM = 3$. Therefore, $ON = 3$.

Segments of Chords, Secants, and Tangents

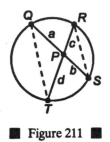

■ Figure 211 ■

In Figure 211, chords $\overline{QS}$ and $\overline{RT}$ intersect at P. By drawing $\overline{QT}$ and $\overline{RS}$, it can be proven that $\triangle QPT \sim \triangle RPS$. Since the ratios of corresponding sides of similar triangles are equal, $a/c = d/b$. The *Cross-Products Property* produces $(a)(b) = (c)(d)$. This is stated as a theorem.

Theorem 83: If two chords intersect inside a circle, then the product of the segments of one chord equals the product of the segments of the other chord.

Example 18: Find x in each of the following figures.

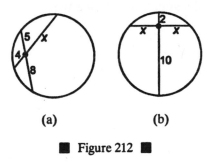

(a)   (b)

■ Figure 212 ■

(a) By *Theorem 83,* $4(x) = (5)(8)$

$$4x = 40$$

$$x = 10$$

(b) By *Theorem 83,* $(x)(x) = (2)(10)$

$$x^2 = 20$$

$$x = \sqrt{20}$$

$$x = \sqrt{4}\sqrt{5}$$

$$x = 2\sqrt{5}$$

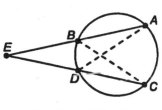

■ Figure 213 ■

In Figure 213, secant segments $\overline{AB}$ and $\overline{CD}$ intersect outside the circle at E. By drawing $\overline{BC}$ and $\overline{AO}$, it can be proven that $\triangle EBC \sim \triangle EDA$. This makes

$$\frac{EB}{ED} = \frac{EC}{EA}$$

By using the *Cross-Products Property,*

$$(EB)(EA) = (ED)(EC)$$

This is stated as a theorem.

Theorem 84: If two secant segments intersect outside a circle, then the product of the secant segment with its external portion equals the product of the other secant segment with its external portion.

Example 19: Find x in each of the following figures.

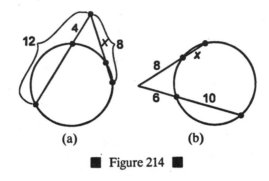

■ Figure 214 ■

(a) By *Theorem 84,* $8(x) = (12)(4)$

$$8x = 48$$

$$x = 6$$

(b) By *Theorem 84,* $(x + 8)(8) = (10 + 6)(6)$

$$8x + 64 = (16)(6)$$

$$8x + 64 = 96$$

$$8x = 32$$

$$x = 4$$

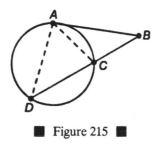

■ Figure 215 ■

In Figure 215, tangent segment $\overline{AB}$ and secant segment $\overline{BD}$ intersect outside the circle at B. By drawing $\overline{AC}$ and $\overline{AD}$, it can be proven that $\triangle ADB \sim \triangle CAB$. Therefore,

$$\frac{AB}{BC} = \frac{BD}{AB}$$

Applying the *Cross-Products Property*,

$$(AB)^2 = (BD)(BC)$$

This is stated as a theorem.

Theorem 85: If a tangent segment and a secant segment intersect outside a circle, then the square of the measure of the tangent segment equals the product of the measures of the secant segment and its external portion.

Also,

Theorem 86: If two tangent segments intersect outside a circle, then the tangent segments have equal measures.

Example 20: Find x in the following figures.

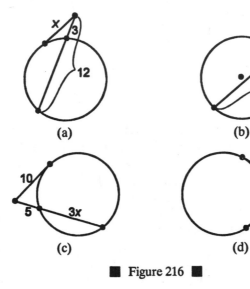

(a) (b)

(c) (d)

■ Figure 216 ■

(a) By *Theorem 85,* $\quad x^2 = (12)(3)$

$$x^2 = 36$$
$$x = \sqrt{36}$$
$$x = 6$$

(b) By *Theorem 85,* $\quad 12^2 = (9x)(x)$

$$144 = 9x^2$$
$$16 = x^2$$
$$\sqrt{16} = x$$
$$4 = x$$

(c) By *Theorem 85*, $10^2 = (3x + 5)(5)$

$$100 = 15x + 25$$
$$75 = 15x$$
$$5 = x$$

(d) By *Theorem 86*, $x = 13$

Arc Length and Sectors

Arc length. As mentioned earlier in this section, an arc can be measured either in degrees or in unit length. In Figure 217, $l\,\widehat{AB}$ is a portion of the circumference of the circle. The portion is determined by the size of its corresponding central angle. A proportion will be created that compares a portion of the circle to the whole circle first in degree measure and then in unit length.

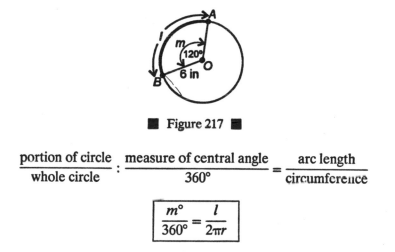

■ Figure 217 ■

$$\frac{\text{portion of circle}}{\text{whole circle}} : \frac{\text{measure of central angle}}{360°} = \frac{\text{arc length}}{\text{circumference}}$$

$$\boxed{\frac{m°}{360°} = \frac{l}{2\pi r}}$$

With the use of this propotion, $l\,\widehat{AB}$ can now be found. In Figure 217, the measure of the central angle = 120°, circumference = $2\pi r$, and r = 6 inches.

$$\frac{120°}{360°} = \frac{l\,\widehat{AB}}{12\pi \text{ inches}}$$

Reduce 120°/360° to 1/3.

$$\frac{1}{3} = \frac{l\,\widehat{AB}}{12\pi \text{ inches}}$$

$$3(l\,\widehat{AB}) = 12\pi \text{ inches}$$

$$l\,\widehat{AB} = 4\pi \text{ inches}$$

Example 21: In Figure 218, $l\,\widehat{AB} = 8\pi$ inches. The radius of the circle is 16 inches. Find $m \angle AOB$.

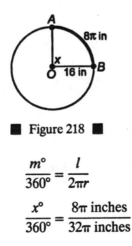

■ Figure 218 ■

$$\frac{m°}{360°} = \frac{l}{2\pi r}$$

$$\frac{x°}{360°} = \frac{8\pi \text{ inches}}{32\pi \text{ inches}}$$

Reduce $8\pi/32\pi$ to $1/4$.

$$\frac{x}{360} = \frac{1}{4}$$

$$4x = 360°$$

$$x = 90°$$

So, $m \angle AOB = 90°$.

Sector of a circle. A **sector of a circle** is a region bounded by two radii and an arc of the circle.

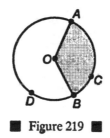

■ Figure 219 ■

In Figure 219, $OACB$ is a sector. $\overarc{ACB}$ is the arc of sector $OACB$. $OADB$ is also a sector. $\overarc{ADB}$ is the arc of sector $OADB$. The area of a sector is a portion of the entire area of the circle. This can be expressed as a proportion.

$$\frac{\text{portion of circle}}{\text{whole circle}} : \frac{\text{measure of central angle}}{360°} = \frac{\text{area of sector}}{\text{area of circle}}$$

$$\boxed{\frac{m°}{360°} = \frac{\text{area of sector}}{\pi r^2}}$$

Example 22: In Figure 220, find the area of sector *OACB*.

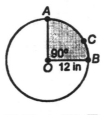

■ Figure 220 ■

$$\frac{\text{measure of central angle}}{360°} = \frac{\text{area of sector}}{\text{area of circle}}$$

$$\frac{90}{360} = \frac{\text{area of sector}}{\pi(12)^2 \text{ in}^2} \quad (\text{area of a circle} = \pi r^2)$$

Reduce 90°/360° to 1/4.

$$\frac{1}{4} = \frac{\text{area of sector}}{144\pi \text{ in}^2}$$

$$4(\text{area sector } OACB) = 144\pi \text{ in}^2$$

$$\text{area sector } OACB = 36\pi \text{ in}^2$$

Example 23: In Figure 221, find the area of sector *RQTS*.

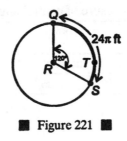

■ Figure 221 ■

$$\frac{m°}{360°} = \frac{l\,\widehat{QS}}{2\pi r}$$

$$\frac{120}{360} = \frac{24\pi}{2\pi r}$$

Reduce 120/360 to 1/3 and $24\pi/2\pi r$ to $12/r$.

$$\frac{1}{3} = \frac{12}{r}$$

$$r = 36$$

The radius of this circle is 36 ft, so the area of the circle is $\pi(36)^2$ or 1296π ft². Therefore,

$$\frac{m°}{360°} = \frac{\text{area of sector}}{\text{area of circle}}$$

$$\frac{120}{360} = \frac{\text{area of sector}}{1296\pi \text{ ft}^2}$$

Reduce 120/360 to 1/3.

$$\frac{1}{3} = \frac{\text{area of sector}}{1296\pi \text{ ft}^2}$$

$$3(\text{area sector } RQTS) = 1296\pi \text{ ft}^2$$

$$\text{area sector } RQTS = 432\pi \text{ ft}^2$$

Summary of Angle, Segment, Arc Length, and Sector Relationships

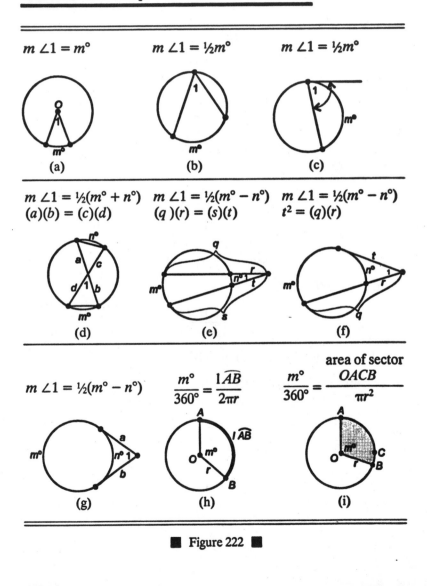

$m \angle 1 = m°$ $m \angle 1 = \frac{1}{2}m°$ $m \angle 1 = \frac{1}{2}m°$

(a)   (b)   (c)

$m \angle 1 = \frac{1}{2}(m° + n°)$ $m \angle 1 = \frac{1}{2}(m° - n°)$ $m \angle 1 = \frac{1}{2}(m° - n°)$
$(a)(b) = (c)(d)$ $(q)(r) = (s)(t)$ $t^2 = (q)(r)$

(d)   (e)   (f)

$m \angle 1 = \frac{1}{2}(m° - n°)$ $\dfrac{m°}{360°} = \dfrac{l\,\overset{\frown}{AB}}{2\pi r}$ $\dfrac{m°}{360°} = \dfrac{\text{area of sector } OACB}{\pi r^2}$

(g)   (h)   (i)

■ Figure 222 ■

Prisms

A **prism** is a type of solid that has the following characteristics.

- **Bases.** A prism has two **bases** which are congruent polygons lying in parallel planes.

- **Lateral edges.** The lines formed by connecting the corre sponding vertices, which forms a series of parallel segments, are **lateral edges.**

- **Lateral faces.** The parallelograms formed by the lateral edges are **lateral faces.**

A prism is named by the polygon that is its base.

- **Altitude.** An **altitude** of a prism is a segment perpendicular to the planes of the bases with an endpoint in each plane.

- **Oblique prism.** An **oblique prism** is a prism whose lateral edges are not perpendicular to the base.

- **Right prism.** A **right prism** is a prism whose lateral edges are perpendicular to the bases. In a right prism, a lateral edge is also an altitude.

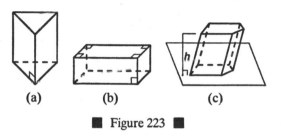

(a)   (b)   (c)

■ Figure 223 ■

In Figure 223, prism (a) is a right triangular prism, prism (b) is a right rectangular prism, and prism (c) is an oblique pentagonal prism. The altitude in prism (c) is called *h*.

Right Prisms

Lateral area of a right prism. The **lateral area of a right prism** is the sum of the areas of all the lateral faces.

Theorem 87: The lateral area, *L.A.*, of a right prism of altitude *h* and perimeter *p* is given by the following equation.

$$L.A._{\text{right prism}} = (p)(h) \text{ units}^2$$

Example 1: Find the lateral area of the right hexagonal prism in Figure 224.

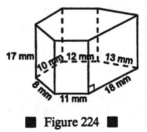

17 mm   10 mm   12 mm   13 mm   8 mm   11 mm   18 mm

■ Figure 224 ■

$$L.A._{\text{right prism}} = (p)(h) \text{ units}^2$$
$$= (11 + 18 + 13 + 12 + 10 + 8)(17) \text{ mm}^2$$
$$= (72)(17) \text{ mm}^2$$
$$= 1224 \text{ mm}^2$$

Total area of a right prism. The **total area of a right prism** is the sum of the lateral area and the areas of the two bases. Since the bases are congruent, their areas are equal.

Theorem 88: The total area, *T.A.*, of a right prism with lateral area *L.A.* and a base area *B* is given by the following equation.

$$T.A._{\text{right prism}} = L.A. + 2B \qquad \text{or} \qquad T.A._{\text{right prism}} = (p)(h) + 2B$$

Example 2: Find the total area of the triangular prism in Figure 225.

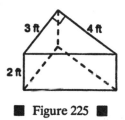

■ Figure 225 ■

The base of this prism is a right triangle with legs of 3 ft and 4 ft.

$\text{hypotenuse}^2 = 3^2 + 4^2$ (*Pythagorean Theorem*)

$\text{hypotenuse}^2 = 9 + 16$

$\text{hypotenuse}^2 = 25$

$\text{hypotenuse} = \sqrt{25}$

$\text{hypotenuse} = 5 \text{ ft}$

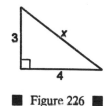

■ Figure 226 ■

The perimeter of the base is (3 + 4 + 5) ft, or 12 ft.

Since the triangle is a right triangle, its legs can be used as base and height of the triangle.

$$Area_{\text{triangle}} = \tfrac{1}{2}(b)(h)$$
$$= \tfrac{1}{2}(3)(4) \text{ ft}^2$$
$$= 6 \text{ ft}^2$$

The altitude of the prism is given as 2 ft. Therefore,

$$T.A._{\text{right prism}} = L.A. + 2B \text{ units}^2$$
$$= (p)(h) + 2B \text{ units}^2$$
$$= (12)(2) + (2)(6) \text{ ft}^2$$
$$= 24 + 12 \text{ ft}^2$$
$$= 36 \text{ ft}^2$$

Interior space of a solid. Lateral area and total area are measurements of the surface of a solid. The **interior space of a solid** can also be measured.

■ **Cube.** A **cube** is a square right prism whose lateral edges are the same length as a side of the base. Figure 227 is a cube.

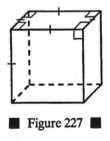

■ Figure 227 ■

■ **Volume.** The **volume** of a solid is the number of cubes necessary to entirely fill the interior of the solid. In Figure 228, the right rectangular prism measures 3 inches by 4 inches by 5 inches.

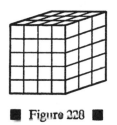

■ Figure 228 ■

This prism could be filled with cubes 1 inch on each side, called a **cubic inch.** The top layer would have 12 such cubes. Since there are 5 such layers, it takes 60 of these cubes to fill this solid. The volume of this prism is then 60 cubic inches.

Theorem 89: The volume, V, of a right prism with a base area B and an altitude h is given by the following equation.

$$V_{\text{right prism}} = (B)(h) \text{ units}^2$$

Example 3: Figure 229 is an isosceles trapezoidal right prism. Find (a) *L.A.*, (b) *T.A.*, and (c) *V*.

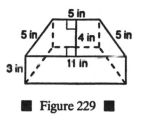

■ Figure 229 ■

(a) $L.A._{\text{right prism}} = (p)(h)$ units^2

(Note: the h refers to the altitude of the prism, not the height of the trapezoid.)

$$= (5 + 5 + 5 + 11)(3) \text{ in}^2$$
$$= (26)(3) \text{ in}^2$$
$$= 78 \text{ in}^2$$

(b) $B = Area_{\text{trapezoid}}$ $T.A._{\text{right prism}} = L.A. + 2B$ units^2

$B = \frac{1}{2}(5 + 11)(4) \text{ in}^2$ $= 78 + 2(32)$

$B = \frac{1}{2}(16)(4) \text{ in}^2$ $= 78 + 64$

$B = 32 \text{ in}^2$ $= 142 \text{ in}^2$

(c) $V_{\text{right prism}} = (B)(h)$ units^3

(Note: The h refers to the altitude of the prism, not the height of the trapezoid.)

$$= (32)(3) \text{ in}^3$$
$$= 96 \text{ in}^3$$

Right Circular Cylinders

A prism-shaped solid whose bases are circles is a **cylinder.** If the segment joining the centers of the circles of a cylinder is perpendicular to the planes of the bases, then the cylinder is a **right circular cylinder.** In Figure 230, cylinder (a) is a right circular cylinder and cylinder (b) is an oblique circular cylinder.

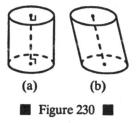

(a)   (b)

■ Figure 230 ■

Lateral area, total area, and volume for right circular cylinders are found in the same way as they are for right prisms.

If a cylinder is pictured as a soup can, then its lateral area would be the area of the label. If the label is carefully peeled off, it would appear as in Figure 231.

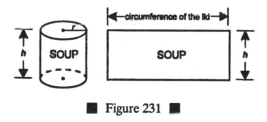

■ Figure 231 ■

The area of the label becomes the area of a rectangle with a height the same as the altitude of the can and a base the same as the circumference of the lid of the can.

Theorem 90: The lateral area, *L.A.*, of a right circular cylinder with a base circumference C and an altitude h is given by the following equation.

$$L.A._{\text{right circular cylinder}} = (C)(h) \text{ units}^2$$

$$= (2\pi r)(h) \text{ units}^2$$

Theorem 91: The total area, *T.A.*, of a right circular cylinder with lateral area *L.A.* and a base area B is given by the following equation.

$$T.A._{\text{right circular cylinder}} = L.A. + 2B \text{ units}^2$$

$$= (2\pi r)(h) + 2\pi r^2 \text{ units}^2$$

$$= 2\pi r(h + r) \text{ units}^2$$

Theorem 92: The volume of a right circular cylinder, V, with a base area B and altitude h is given by the following equation.

$$V_{\text{right circular cylinder}} = (B)(h) \text{ units}^3$$

$$= (\pi r^2)(h) \text{ units}^3$$

Example 4: Use Figure 232 of a right circular cylinder to find (a) *L.A.*, (b) *T.A.*, and (c) *V*.

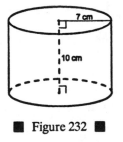

■ Figure 232 ■

(a) $L.A._{\text{right circular cylinder}} = (C)(h)$ units^2

$= (2)(\pi)(7)(10)$ cm^2

$= 140\pi$ cm^2

(b) $T.A._{\text{right circular cylinder}} = L.A. + 2B$ units^2

$= 140\pi + 2(\pi)(7)^2$ cm^2

$= 140\pi + 98\pi$ cm^2

$= 238\pi$ cm^2

(c) $V_{\text{right circular cylinder}} = (B)(h)$ units^3

$= (\pi)(7)^2(10)$ cm^3

$= (49\pi)(10)$ cm^3

$= 490\pi$ cm^3

Pyramids

A **pyramid** is a solid that has the following characteristics.

- It has one **base,** which is a polygon.
- The vertices of the base are each joined to a point, not in the plane of the base, called the **vertex** of the pyramid.
- The triangular sides, which each meet the vertex, are its **lateral faces.**
- The segments where the lateral faces intersect are **lateral edges.**
- The perpendicular segment from the vertex to the plane of the base is the **altitude of the pyramid.**

Regular Pyramids

A **regular pyramid** is a pyramid whose base is a regular polygon, and its lateral edges are all equal in length. A pyramid is named by its base. Figure 233 shows some examples of regular pyramids.

regular
triangular pyramid

regular
square pyramid

regular
hexagonal pyramid

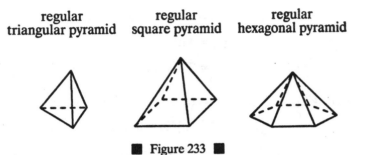

■ Figure 233 ■

The lateral faces of a regular pyramid are congruent isosceles triangles. The altitude of any of these triangles is the **slant height** of the regular pyramid. Pictured in Figure 234 is a square pyramid.

Square *ABCD* is its base.

V is the vertex.

Triangle *VAB* is a lateral face.

VA is a lateral edge.

h is the altitude.

l is the slant height.

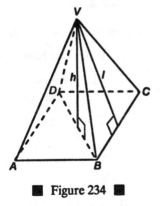

■ Figure 234 ■

Pyramids also have a lateral area, total area, and volume.

Theorem 93: The lateral area, *L.A.*, of a regular pyramid with slant height *l* and base perimeter *p* is given by the following equation.

$$L.A._{\text{regular pyramid}} = \tfrac{1}{2}(p)(l) \text{ units}^2$$

Example 5: Find the lateral area of the square pyramid in Figure 235.

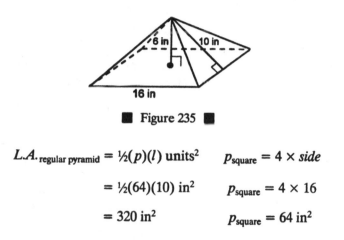

■ Figure 235 ■

$$L.A._{\text{regular pyramid}} = \tfrac{1}{2}(p)(l) \text{ units}^2 \qquad p_{\text{square}} = 4 \times side$$

$$= \tfrac{1}{2}(64)(10) \text{ in}^2 \qquad p_{\text{square}} = 4 \times 16$$

$$= 320 \text{ in}^2 \qquad p_{\text{square}} = 64 \text{ in}^2$$

Since a pyramid has only one base, its total area is the sum of the lateral area and the area of its base.

Theorem 94: The total area, *T.A.*, of a regular pyramid with lateral area *L.A.* and base area *B* is given by the following equation.

$$T.A._{\text{regular pyramid}} = L.A. + B \text{ units}^2$$

$$= \tfrac{1}{2}(p)(l) + B \text{ units}^2$$

Example 6: Find the total area of the regular pyramid in Figure 235.

The base of the regular pyramid is a square. $A_{\text{square}} = (side)^2$. Therefore, $B = 16^2$ in^2, or $B = 256$ in^2.

$$T.A._{\text{regular pyramid}} = L.A. + B \text{ units}^2$$

From the previous example, $L.A. = 320$ in^2.

$$= 320 + 256 \text{ in}^2$$

$$= 576 \text{ in}^2$$

Theorem 95: The volume, V, of a regular pyramid with base area B and altitude h is given by the following equation.

$$V_{\text{regular pyramid}} = \tfrac{1}{3}(B)(h) \text{ units}^3$$

Example 7: Find the volume of the regular pyramid in Figure 235.

From the previous example, $B = 256$ in^2. The figure indicates that $h = 6$ in.

$$V_{\text{regular pyramid}} = \tfrac{1}{3}(B)(h) \text{ units}^3$$

$$= \tfrac{1}{3}(256)(6) \text{ in}^3$$

$$= 512 \text{ in}^3$$

Right Circular Cones

A **right circular cone** is similar to a regular pyramid except that its base is a circle. The vocabulary and equations pertaining to the right circular cone are similar to those for the regular pyramid. Refer to Figure 236 for the vocabulary regarding right circular cones.

right circular cone

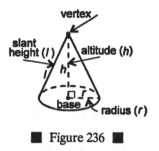

■ Figure 236 ■

Theorem 96: The lateral area, *L.A.*, of a right circular cone with base circumference C and slant height l is given by the following equation.

$$L.A._{\text{right circular cone}} = \tfrac{1}{2}(C)(l) \text{ units}^2$$
$$= \tfrac{1}{2}(2\pi)(r)(l) \text{ units}^2$$
$$= \pi r l \text{ units}^2$$

Theorem 97: The total area, *T.A.*, of a right circular cone with lateral area *L.A.* and base area B is given by the following equation.

$$T.A._{\text{right circular cone}} = L.A. + B \text{ units}^2$$
$$= \pi r l + \pi r^2 \text{ units}^2$$
$$= \pi r(l + r) \text{ units}^2$$

Theorem 98: The volume, V, of a right circular cone with base area B and altitude h is given by the following equation.

$$V_{\text{right circular cone}} = \tfrac{1}{3}(B)(h) \text{ units}^3$$
$$= \tfrac{1}{3}(\pi r^2)(h) \text{ units}^3$$

Example 8: Use Figure 237 of a right circular cone to find (a) *L.A.*, (b) *T.A.*, and (c) *V.*

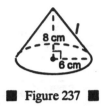

■ Figure 237 ■

(a) The slant height, radius, and altitude of a right circular cone form a right triangle as pictured in Figure 238.

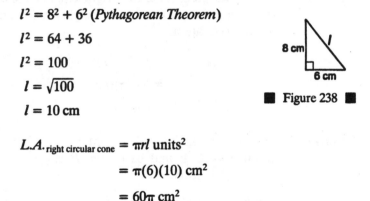

$l^2 = 8^2 + 6^2$ (*Pythagorean Theorem*)

$l^2 = 64 + 36$

$l^2 = 100$

$l = \sqrt{100}$

$l = 10$ cm

■ Figure 238 ■

$L.A._{\text{right circular cone}} = \pi r l \text{ units}^2$

$= \pi(6)(10) \text{ cm}^2$

$= 60\pi \text{ cm}^2$

(b) $T.A._{\text{right circular cone}} = L.A. + B \text{ units}^2$

$= 60\pi + \pi(6)^2 \text{ cm}^2$

$= 60\pi + 36\pi \text{ cm}^2$

$= 96\pi \text{ cm}^2$

(c) $V_{\text{right circular cone}} = \frac{1}{3}(B)(h)$ units^3

$= \frac{1}{3}(36\pi)(8)$ cm^3

$= 96\pi$ cm^3

Spheres

A **sphere** is the set of all points in space that are at a given distance from a given point. That distance is the **radius of the sphere.** Since a sphere has no bases, its area is referred to as its **surface area.**

Theorem 99: The surface area, *S*, of a sphere with radius *r* is given by the following equation.

$$S_{\text{sphere}} = 4\pi r^2 \text{ units}^2$$

Theorem 100: The volume of a sphere, *V*, with radius *r* is given by the following equation.

$$V_{\text{sphere}} = \frac{4}{3}\pi r^3 \text{ units}^3$$

Example 9: Figure 239 represents a sphere with radius *r*. If *r* = 9 cm, find (a) *S* and (b) *V*.

■ Figure 239 ■

(a)　$S_{sphere} = 4\pi r^2$ units^2

$= 4\pi(9)^2$ cm^2

$= 4\pi(81)$ cm^2

$= 324\pi$ cm^2

(b)　$V_{sphere} = \frac{4}{3}\pi r^3$ units^3

$= \frac{4}{3}\pi(9)^3$ cm^3

$= \frac{4}{3}\pi(729)$ cm^3

$= 972\pi$ cm^3

Summary of Equations Concerning Geometric Solids

C = circumference of a base
r = radius of a circle
p = perimeter of a base
B = area of a base
h = altitude
l = slant height
S = surface area of a sphere

Name	Example Figure	Lateral Area	Total Area	Volume
right prism	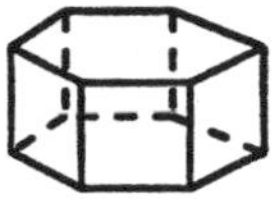 (a)	ph	$L.A. + 2B$ $= ph + 2B$	Bh
right circular cylinder	 (b)	$Ch = 2\pi rh$	$L.A. + 2B$ $= 2\pi rh + 2\pi r^2$ $= 2\pi r(h + r)$	Bh
regular pyramid	 (c)	$\frac{1}{2}pl$	$L.A. + B$ $= \frac{1}{2}pl + B$	$\frac{1}{3}Bh$
right circular cone	 (d)	$\frac{1}{2}Cl$ $= \frac{1}{2}(2\pi r)l$ $= \pi rl$	$L.A. + B$ $= \pi rl + \pi r^2$ $= \pi r(l + r)$	$\frac{1}{3}Bh$ $= \frac{1}{3}\pi r^2 h$
sphere	 (e)	none	$S = 4\pi r^2$	$\frac{4}{3}\pi r^3$

■ Figure 240 ■

Points and Coordinates

- **Coordinates of a point.** Each point on a number line is assigned a number. In the same way, each point in a plane is assigned a pair of numbers called the **coordinates of the point.**

- **x-axis and y-axis.** To locate points in a plane, two perpendicular lines are used: a horizontal line called the **x-axis** and a vertical line called the **y-axis.**

- **Origin.** The point of intersection of the x-axis and y-axis is called the **origin.**

- **Coordinate plane.** The x-axis, y-axis, and all the points in their plane are called a **coordinate plane.**

- **Ordered pairs.** Every point in a coordinate plane is named by a pair of numbers whose order is important. This pair of numbers, written in parentheses and separated by a comma, is the **ordered pair** for the point.

- **x-coordinate.** The number to the left of the comma in an ordered pair is the **x-coordinate** of the point and indicates the amount of movement along the x-axis from the origin. The movement is to the right if the number is positive and to the left if the number is negative.

- **y-coordinate.** The number to the right of the comma in an ordered pair is the **y-coordinate** of the point and indicates the amount of movement perpendicular to the x-axis. The movement is above the x-axis if the number is positive **and** below the x-axis if the number is negative.

The ordered pair for the origin is (0, 0).

The x-axis and y-axis separate the coordinate plane into four regions called **quadrants.** The upper right quadrant is quadrant I; the upper left quadrant is quadrant II; the lower left quadrant is quadrant III; and the lower right quadrant is quadrant IV. Notice that

in quadrant I, x is always positive and y is always positive (+, +);
in quadrant II, x is always negative and y is always positive (−, +);
in quadrant III, x is always negative and y is always negative (−, −);
In quadrant IV, x is always positive and y is always negative (+, −).

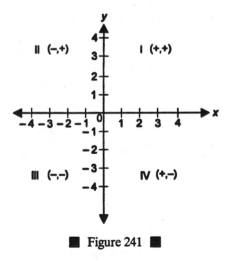

■ Figure 241 ■

The point associated with an ordered pair of real numbers is called the **graph** of the ordered pair.

Example 1: Identify the points *A*, *B*, *C*, *D*, *E*, and *F* on the coordinate graph in Figure 242.

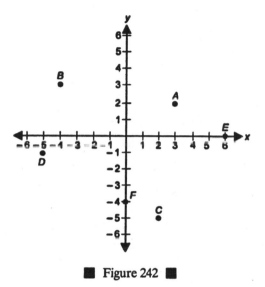

■ Figure 242 ■

$A(3, 2)$ $B(-4, 3)$ $C(2, -5)$ $D(-5, -1)$ $E(6, 0)$ $F(0, -4)$

Example 2: Rectangle *ABCD* has coordinates as follows: $A(-5, 2)$ $B(8, 2)$ $C(8, -4)$. Find the coordinates of *D*.

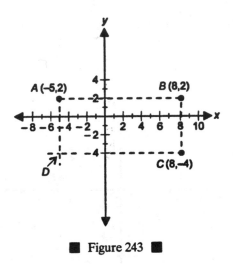

■ Figure 243 ■

A graph is helpful in solving this problem. Refer to Figure 243. The coordinate of D becomes $(-5, -4)$.

Example 3: Use Figure 243 to find the following distances: (a) from A to B (called AB) and (b) from B to C (called BC).

(a) $AB = 8 - (-5)$ and (b) $BC = 2 - (-4)$ (*Postulate 7*)

$AB = 13$ $BC = 6$

Distance Formula

In Figure 244, A is $(2, 2)$, B is $(5, 2)$, and C is $(5, 6)$.

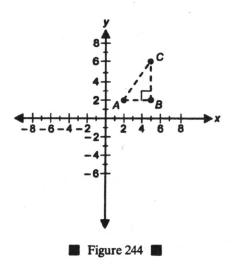

■ Figure 244 ■

To find AB or BC, only simple subtracting is necessary

$$AB = 5 - 2 \quad \text{and} \quad BC = 6 - 2$$
$$AB = 3 \qquad\qquad\qquad BC = 4$$

To find AC, though, simply subtracting is not sufficient. Triangle ABC is a right triangle with $\overline{AC}$ the hypotenuse. Therefore, by the *Pythagorean Theorem*,

$$AC^2 = AB^2 + BC^2$$
$$AC = \sqrt{AB^2 + BC^2}$$
$$AC = \sqrt{3^2 + 4^2}$$
$$AC = \sqrt{9 + 16}$$
$$AC = \sqrt{25}$$
$$AC = 5$$

If A is represented by the ordered pair (x_1, y_1) and C is represented by the ordered pair (x_2, y_2), then $AB = (x_2 - x_1)$ and $BC = (y_2 - y_1)$.

Then

$$AC = \sqrt{(x_2 - x_1)^2 + (y_2 - y_1)^2}$$

This is stated as a theorem.

Theorem 101: If the coordinates of two points are (x_1, y_1) and (x_2, y_2), then the distance between them, d, is given by the following formula (*Distance Formula*).

$$d = \sqrt{(x_2 - x_1)^2 + (y_2 - y_1)^2}$$

Example 4: Use the *Distance Formula* to find the distance between the points with coordinates $(-3, 4)$ and $(5, 2)$.

Let $(-3, 4) = (x_1, y_1)$ and $(5, 2) = (x_2, y_2)$. Then

$$d = \sqrt{(5 - -3)^2 + (2 - 4)^2}$$
$$d = \sqrt{(8)^2 + (-2)^2}$$
$$d = \sqrt{64 + 4}$$
$$d = \sqrt{68}$$
$$d = \sqrt{4}\sqrt{17}$$
$$d = 2\sqrt{17}$$

Example 5: A triangle has vertices $A(12, 5)$, $B(5, 3)$, and $C(12, 1)$. Show that the triangle is isosceles.

By the *Distance Formula,*

$$AB = \sqrt{(5 - 12)^2 + (3 - 5)^2} \qquad BC = \sqrt{(12 - 5)^2 + (1 - 3)^2}$$
$$AB = \sqrt{(-7)^2 + (-2)^2} \qquad BC = \sqrt{7^2 + (-2)^2}$$

$$AB = \sqrt{49 + 4} \qquad\qquad BC = \sqrt{49 + 4}$$

$$AB = \sqrt{53} \qquad\qquad\quad BC = \sqrt{53}$$

Since $AB = BC$, triangle ABC is isosceles.

Midpoint Formula

Numerically, the midpoint of a segment can be considered to be the average of its endpoints. This concept helps in remembering a formula for finding the midpoint of a segment given the coordinates of its endpoints. Recall that the average of two numbers is found by dividing their sum by two.

Theorem 102: If the coordinates of A and B are (x_1, y_1) and (x_2, y_2) respectively, then the midpoint, M, of AB is given by the following formula (*Midpoint Formula*).

$$M = \left(\frac{x_1 + x_2}{2}, \frac{y_1 + y_2}{2}\right)$$

Example 6: In Figure 245, R is the midpoint between $Q(-9, -1)$ and $T(-3, 7)$. Find its coordinate and use the *Distance Formula* to verify that it is in fact the midpoint of $\overline{QT}$.

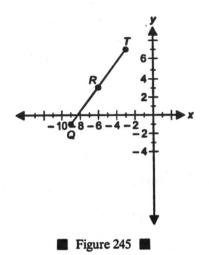

■ Figure 245 ■

By the *Midpoint Formula,*

$$R = \left(\frac{-9 + -3}{2}, \frac{-1 + 7}{2} \right)$$

$$R = \left(\frac{-12}{2}, \frac{6}{2} \right)$$

$$R = (-6, 3)$$

By the *Distance Formula,*

$$QR = \sqrt{(-6 - -9)^2 + (3 - -1)^2} \qquad TR = \sqrt{(-6 - -3)^2 + (3 - 7)^2}$$

$$QR = \sqrt{3^2 + 4^2} \qquad\qquad\qquad TR = \sqrt{(-3)^2 + (-4)^2}$$

$$QR = \sqrt{9 + 16} \qquad\qquad\qquad TR = \sqrt{9 + 16}$$

$$QR = \sqrt{25} \qquad\qquad\qquad\quad TR = \sqrt{25}$$

$$QR = 5 \qquad\qquad\qquad\qquad\quad TR = 5$$

Since $QR = TR$ and Q, T, and R are collinear, then R is the midpoint of $\overline{QT}$

Example 7: If the midpoint of $\overline{AB}$ is $(-3, 8)$ and A is $(12, -1)$, find the coordinates of B.

Let the coordinates of B be (x, y). Then by the *Midpoint Formula*,

$$(-3, 8) = \left(\frac{12 + x}{2}, \frac{-1 + y}{2} \right)$$

$$-3 = \frac{12 + x}{2} \quad \text{and} \quad 8 = \frac{-1 + y}{2}$$

Multiply each side of each equation by 2.

$$-6 = 12 + x \quad \text{and} \quad 16 = -1 + y$$
$$-18 = x \quad \text{and} \quad 17 = y$$

Therefore, the coordinates of B are $(-18, 17)$.

Slope of a Line

The **slope of a line** is a measurement of the steepness and direction of a nonvertical line. When a line rises from left to right, the slope is a positive number. Figure 246(a) shows a line with a positive slope. When a line falls from left to right, the slope is a negative number. Figure 246(b) shows a line with a negative slope. The x-axis or any line parallel to the x-axis has a slope of zero. Figure 246(c) shows a line whose slope is zero. The y-axis or any line parallel to the y-axis has no defined slope. Figure 246(d) shows a line with an undefined slope.

a line with a positive slope   a line with a negative slope

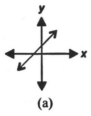 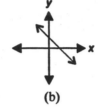

(a)   (b)

a line with a zero slope   a line with an undefined slope

(c)   (d)

■ Figure 246 ■

If m represents the slope of a line and A and B are points with coordinates (x_1, y_1) and (x_2, y_2) respectively, then the slope of the line passing through A and B is given by the following formula.

$$m = \frac{y_2 - y_1}{x_2 - x_1}, x_2 \neq x_1$$

Since A and B cannot be points on a vertical line, x_1 and x_2 cannot be equal to one another. If $x_1 = x_2$, then the line is vertical and the slope is undefined.

Example 8: Use Figure 247 to find the slopes of lines, $a, b, c,$ and d.

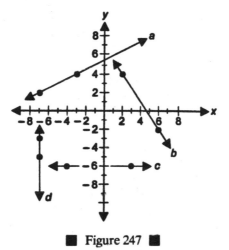

■ Figure 247 ■

(a) Line a passes through the points $(-7, 2)$ and $(-3, 4)$.

$$m = \frac{4-2}{-3--7}$$

$$m = \frac{2}{4}$$

$$m = \frac{1}{2}$$

(b) Line b passes through the points $(2, 4)$ and $(6, -2)$.

$$m = \frac{-2-4}{6-2}$$

$$m = \frac{-6}{4}$$

$$m = -\frac{3}{2}$$

(c) Line c is parallel to the x-axis. Therefore, $m = 0$.

(d) Line d is parallel to the y-axis. Therefore, line d has an undefined slope.

Example 9: A line passes through $(-5, 8)$ with a slope of 2/3. If another point on this line has coordinates $(x, 12)$, find x.

$$m = \frac{y_2 - y_1}{x_2 - x_1}$$

$$\frac{2}{3} = \frac{12 - 8}{x - -5}$$

$$\frac{2}{3} = \frac{4}{x + 5}$$

$$2(x + 5) = 4(3) \quad \text{(Cross-Products Property)}$$

$$2x + 10 = 12$$

$$2x = 2$$

$$x = 1$$

Slopes of Parallel and Perpendicular Lines

If lines are parallel, they slant in exactly the same directions. If they are nonvertical, their steepness is exactly the same.

Theorem 103: If two nonvertical lines are parallel, then they have the same slope.

Theorem 104: If two lines have the same slope, then the lines are nonvertical parallel lines.

If two lines are perpendicular and neither one is vertical, then one of the lines has a positive slope and the other has a negative slope. Also, the absolute values of their slopes are reciprocals.

Theorem 105: If two nonvertical lines are perpendicular, then their slopes are opposite reciprocals of one another, or the product of their slopes is −1.

Theorem 106: If the slopes of two lines are opposite reciprocals of one another, or the product of their slopes is −1, then the lines are nonvertical perpendicular lines.

Since horizontal and vertical lines are always perpendicular, then lines having a zero slope and an undefined slope are perpendicular.

Example 10: If line *l* has slope 3/4, then (a) any line parallel to *l* has slope __, and (b) any line perpendicular to *l* has slope __.

(a) 3/4 (*Theorem 103*)

(b) −4/3 (*Theorem 105*)

Example 11: Given points *Q*, *R*, *S*, and *T*, tell which sides, if any, of quadrilateral *QRST* in Figure 248 are parallel or perpendicular.

$$Q(-1, 0) \quad R(1, 1) \quad S(0, 3) \quad T(-3, 4)$$

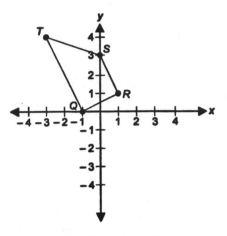

■ Figure 248 ■

$$m_{\overline{QR}} = \frac{1-0}{1--1} \qquad m_{\overline{QR}} = \frac{1}{2}$$

$$m_{\overline{RS}} = \frac{3-1}{0-1} \qquad m_{\overline{RS}} = -\frac{2}{1}$$

$$m_{\overline{ST}} = \frac{4-3}{-3-0} \qquad m_{\overline{ST}} = -\frac{1}{3}$$

$$m_{\overline{QT}} = \frac{4-0}{-3--1} \qquad m_{\overline{QT}} = -\frac{2}{1}$$

So $\overline{QR} \perp \overline{RS}$ and $\overline{QR} \perp \overline{QT}$ (*Theorem 106*)

and $\overline{RS} \parallel \overline{QT}$ (*Theorem 104*)

Equations of Lines

Equations involving one or two variables can be graphed on any *x-y* coordinate plane. In general, it is true that

1. if a point lies on the graph of an equation, then its coordinates make the equation a true statement, and

2. if the coodinates of a point make an equation a true statement, then the point lies on the graph of the equation.

A **linear equation** is any equation whose graph is a line. All linear equations can be written in the form $Ax + By = C$ where $A, B,$ and C are real numbers and A and B are not both zero. Below are examples of linear equations and their respective A, B, and C values.

$x + y = 0$	$3x - 4y = 9$	$x = -6$	$y = 7$
$A = 1$	$A = 3$	$A = 1$	$A = 0$
$B = 1$	$B = -4$	$B = 0$	$B = 1$
$C = 0$	$C = 9$	$C = -6$	$C = 7$

This form for equations of lines is known as the **standard form** for the equation of a line.

The **x-intercept** of a graph is the point where the graph intersects the x-axis. It always has a y-coordinate of zero. A horizontal line, that is not the x-axis, has no x-intercept.

The **y-intercept** of a graph is the point where the graph intersects the y-axis. It always has an x-coordinate of zero. A vertical line, that is not the y-axis, has no y-intercept.

One way to graph a linear equation is to find solutions by giving a value to one variable and solving the resulting equation for other variables. A minimum of two points is necessary to graph a linear equation.

Example 12: Draw the graph of $2x + 3y = 12$ by finding the x-intercept and the y-intercept.

The x-intercept has a y-coordinate of zero. Substituting zero for y, the resulting equation is $2x + 3(0) = 12$. Now solving for x,

$$2x = 12$$

$$x = 6$$

The x-intercept is at (6, 0), or the x-intercept value is 6.

The y-intercept has an x-coordinate of zero. Substituting zero for x, the resulting equation is $2(0) + 3y = 12$. Now solving for y,

$$3y = 12$$

$$y = 4$$

The y-intercept is at (0, 4), or the y-intercept value is 4.

The line can now be graphed by graphing these two points and drawing the line they determine.

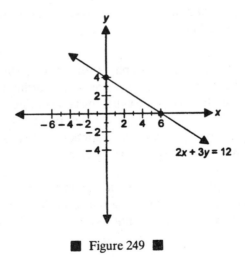

■ Figure 249 ■

Example 13: Draw the graph of $x = 2$.

$x = 2$ is a vertical line whose x-coordinate is always 2.

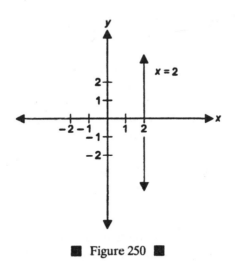

■ Figure 250 ■

Example 14: Draw the graph of $y = -1$.

$y = -1$ is a horizontal line whose y-coordinate is always -1.

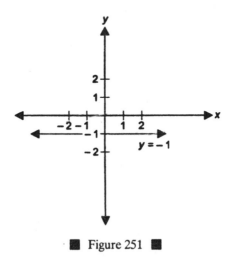

■ Figure 251 ■

Suppose that A is a particular point called (x_1, y_1) and B is any point called (x, y). Then the slope of the line through A and B is represented by

$$\frac{y - y_1}{x - x_1} = m$$

Applying the *Cross-Products Property*, $y - y_1 = m(x - x_1)$. This is the **point-slope form** of a nonvertical line.

Theorem 107: The point-slope form of a line passing through (x_1, y_1) and having slope m is $y - y_1 = m(x - x_1)$.

Example 15: Find the equation of a line containing the points $(-3, 4)$ and $(7, 2)$ and write the equation in (a) point-slope form and (b) standard form.

(a) For the point-slope form, first find the slope, m.

$$m = \frac{2 - 4}{7 - -3}$$

$$m = -\frac{2}{10}$$

$$m = -\frac{1}{5}$$

Now choose either original point—say, $(-3, 4)$.

So, $\boxed{y - 4 = -\tfrac{1}{5}(x - -3) \text{ or } y - 4 = -\tfrac{1}{5}(x + 3)}$

(b) Begin with the point-slope form and clear it of fractions by multiplying both sides by the least common denominator.

$$y - 4 = \tfrac{1}{5}(x + 3)$$

Multiply both sides by 5.

$$5(y - 4) = 5[-\tfrac{1}{5}(x + 3)]$$
$$5y - 20 = -(x + 3)$$
$$5y - 20 = -x - 3$$

Get x and y on one side and the constants on the other side by adding x to both sides and adding 20 to both sides.

$$\boxed{x + 5y = 17}$$

A nonvertical line written in standard form is $Ax + By = C$ with $B \neq 0$. If this equation is solved for y, it becomes

$$By = -Ax + C$$
$$y = -(A/B)x + C/B$$

The value $-(A/B)$ becomes the slope of the line, and C/B becomes the y-intercept value. If $-(A/B)$ is replaced with m and C/B is replaced with b, the equation becomes $y = mx + b$. This is known as the **slope-intercept form** of a nonvertical line.

Theorem 108: The slope-intercept form of a nonvertical line with slope m and y-intercept value b is $y = mx + b$.

Example 16: Find the slope and y-intercept value of the line with equation $3x - 4y = 20$.

Solve $3x - 4y = 20$ for y.

$$-4y = -3x + 20$$
$$y = (\tfrac{3}{4})x - 5$$

Therefore, the slope of the line is 3/4 and the y-intercept value is -5.

Example 17: Line l_1 has equation $2x + 5y = 10$. Line l_2 has equation $4x + 10y = 30$. Line l_3 has equation $15x - 6y = 12$. Which lines, if any, are parallel?

Put each equation into slope-intercept form and determine the slope of each line.

l_1:
$$2x + 5y = 10$$
$$5y = -2x + 10$$
$$y = (-\tfrac{2}{5})x + 2$$
$$\text{slope } l_1 = -\tfrac{2}{5}$$

l_2:
$$4x + 10y = 30$$
$$10y = -4x + 30$$
$$y = (-\tfrac{2}{5})x + 3$$
$$\text{slope } l_2 = -\tfrac{2}{5}$$

l_3:
$$15x - 6y = 12$$
$$-6y = -15x + 12$$
$$y = (\tfrac{5}{2})x - 2$$
$$\text{slope } l_3 = \tfrac{5}{2}$$

Since slope $l_1 = $ slope l_2, then $l_1 \parallel l_2$ by *Theorem 104*.

Since (slope l_1)(slope l_3) $= -1$ and (slope l_2)(slope l_3) $= -1$, then $l_1 \perp l_3$ and $l_2 \perp l_3$ by *Theorem 106*.

Summary of Coordinate Geometry Formulas

If $A(x_1, y_1)$ and $B(x_2, y_2)$, then

distance d, from A to $B =$

$$d = \sqrt{(x_2 - x_1)^2 + (y_2 - y_1)^2}$$

midpoint, M, of $\overline{AB} =$

$$M = \left(\frac{x_1 + x_2}{2}, \frac{y_1 + y_2}{2}\right)$$

slope, m, of $\overleftrightarrow{AB} =$

$$m = \frac{y_2 - y_1}{x_2 - x_1}$$

Equations of lines.

standard form: $Ax + By = C$

$A, B,$ and C are real numbers

A and B are not both zero

point-slope form: $y - y_1 = m(x - x_1)$

(x_1, y_1) is a point on the line and m is the slope of the line

slope-intercept form: $y = mx + b$

m is the slope of the line and b is the y-intercept value

DATE DUE

1/2/16			